Mechanical Engineering

PRINCIPLES OF ARMAMENT DESIGN

H Peter

 www.trafford.com

North America & international
toll-free: 844-688-6899 (USA & Canada)
fax: 812 355 4082

This book was

Electronically type set using Microsoft Word 2000 Professional in 10/12 Century Schoolbook

The graphics drawn in Corel Draw Version 5.00.E2, Corel Corporation 1994

The equations set using Microsoft Equation Editor, Version 3.1, Microsoft Corporation

The workbooks created using Microsoft Excel 2000 Professional

The programmes were run in Matlab Version 6.5.0.180913a Release 13 June 18 2002, The Mathworks Inc

and

The cover designed using Adobe Photoshop CS 1990-2003, Adobe Systems Incorporated

by the author

Acknowledgements

The encouragement from family and friends, especially Solomon Beresa now doing his Masters at Kassel, made this, the second book in my series on armament engineering a reality.

Introduction

The field of armament design is as vast as it is complicated. It is impossible, therefore, to condense into a single volume every design aspect that the armament designer is likely to come across. The contents of this book are aimed at familiarizing the reader with the principles behind the mechanical engineering design of large caliber armament. Based on these principles, it is anticipated that the mechanical engineer will be in a position to approach issues of specific weapon design and produce successful results.

The contents of this book are closely related to the first publication 'Armament Engineering a computer aided approach', also published in collaboration with Trafford. In fact, many of the concepts within are extended from there and many of the equations cross-referenced.

While every effort has been made to keep the contents of this book self-contained, it does presuppose a sound awareness of armament engineering, advanced strength of materials and the design of machine elements on the part of the reader.

Finally, comments and suggestions are welcome and should be addressed to hpeter19@yahoo.com .

The Author

Contents

List of figures   i

Contents of CD ROM   xi

1

Design of Gun Barrels

1.1	General Considerations for Gun Barrel Design	1
1.2	Theories of Failure of Gun Barrel Materials	6
1.3	Conventions used in Gun Barrel Design	13
1.4	Monobloc Gun Barrel Design	17
1.5	Design of a Monobloc Gun Barrel	19
1.6	Autofrettaged Gun Barrel Design	32
1.7	Design of an Autofrettaged Gun Barrel	51
1.8	Rifling Design	64
1.9	Chamber Design	79
1.10	Preliminary Design of a Gun Barrel Chamber	86
1.11	Vibration & Accuracy of Gun Barrels	91
1.12	Vibration Characteristic of a Gun Barrel	100

2

Design of Breech Assemblies

2.1	Design Considerations of Breech Assemblies	108
2.2	Stresses in Interrupted Thread Breech Screws	121
2.3	Estimation of Stresses in an Interrupted Thread Breech Screw	128
2.4	Stresses in Screw Threads of Breech Screws	131
2.5	Semi Automatic Breech Mechanisms	132
2.6	Dynamics of Semi Automatic Breech Operation	134
2.7	Analysis of Dynamics of a Semi Automatic Breech Mechanism	142

3

Recoil Systems

3.1	General Principles of Recoil System Design	147
3.2	Recoil Dynamics	152
3.3	Recoil Computation	165
3.4	Dynamics of Counter Recoil	182
3.5	Counter Recoil Computation	189
3.6	Dynamics of the Buffing Phase	197
3.7	Buffing Phase Computation	204
3.8	Design of Recoil System Components	213

4

Design of Balancing Gears

4.1   Design Considerations of Balancing Gears   221
4.2   Pneumatic Balancing Gears   237
4.3   Design of a Pneumatic Balancing Gear   250
4.4   Spring Type Balancing Gears   263
4.5   Design of a Spring Type Balancing Gear   267

Design of Elevating Mechanisms

5.1   Design Considerations of Elevating Mechanisms   275
5.2   Elevating Gear Trains   279
5.3   Design of a Manually Operated Elevating Gear Train   295
5.4   Design of a Power Operated Elevating Gear Train   300

6

Design of Traversing Mechanisms

6.1	General Considerations for Design of Traversing Mechanisms	311
6.2	Design of Traversing Gear Trains	316
6.3	Design of a Manual Traversing Gear Train	322
6.4	Design of a Power Traversing Gear Train	333

Appendices

A	Useful Conversion Tables	345
B	Moments of Inertia of Sections	351
C	Moments of Inertia of Solids	354
D	Geometric Formulae	358
E	Mechanical Properties of Common Metals	362
F	Mechanical Properties of Popular Carbon & Alloy Steels	363
G	References	365

Index

367

List of Figures

1

1.2

1.2.1   Resolution of combined stresses into three principal directions   6
1.2.2   Strain energy within the limit of proportionality   9
1.2.3   Superimposition of a fourth stress σ on an existing stress system   10

1.3

13.1   Eccentricity of bore and outer surface circles of a gun barrel   15

1.5

1.5.1   Interior ballistics: Shot travel-CMP data   21
1.5.2   Inner dimensions of monobloc gun barrel from the internal ballistics solution   22
1.5.3   76 mm Gun design data sheet   23
1.5.4   Barrel length-CMP, PIMP curves   25
1.5.5   Barrel length-ESP, PIMP, Safety Factor, wall thickness curves   27
1.5.6   Tabulated design data of monobloc gun barrel   28
1.5.7   Stresses in the gun barrel at chamber, mid section and muzzle   31

1.6

1.6.1	Unsupported autofrettage	36
1.6.2	Container autofrettage apparatus	37
1.6.3	Initial clearance between barrel and container before autofrettage	39

1.7

1.7.1	Cross section through a barrel under autofrettage pressure	51
1.7.2	Graphical results of autofrettaged gun barrel design	53
1.7.3	Tabulated results of autofrettaged gun barrel design	54
1.7.4	Yield, working and equivalent stresses in an autofrettaged gun barrel	59

1.8

1.8.1	Internal ballistics worksheet for computation of rifling torque	70
1.8.2	Plot showing rifling torque fir different rifling curves	72
1.8.3	Parameters of rifling profile	74
1.8.4	Leading dimensions of a land	75
1.8.5	Calibre versus depth of grooves, width of lands, number of grooves and width of grooves calculates and actual relations	77
1.8.6	Effect of shape of driving edge of land on torque on driving band	78

1.9

1.9.1	Effect of forcing cone slope	82
1.9.2	Chamber-cartridge case recovery	83
1.9.3	Chamber wall ratio-radial elastic displacement curve	86

1.10

1.10.1	Calibre-cartridge case thickness rough estimation guide	87
1.10.2	Worksheet for chamber design of 76 mm gun (data & formulae)	89
1.10.3	Worksheet for chamber design of 76 mm gun (data & results)	90

1.11

1.11.1	Plot of accuracy-frequency relationship	92
1.11.2	Deflection of a beam by the Area-Moment method	93
1.11.3	Gun barrel as a cantilever with uniform load distribution	95
1.11.4	Deflection of a gun barrel by lumped loading method	98

1.12

1.12.1	Dimensions of a given gun barrel	100
1.12.2	Worksheet for determination of circular speed under static loading	102
1.12.3	Worksheet for 1st iteration for calculation of circular speed under dynamic loading	103
1.12.4	Worksheet for 2nd iteration for calculation of circular speed under dynamic loading	104
1.12.5	Worksheet for 3rd iteration for calculation of circular speed under dynamic loading	105
1.12.6	Worksheet for 4th iteration for calculation of circular speed under dynamic loading, natural frequency & Frequency Ratio	106

2.1

2.1.1	Metal cartridge case and bagged charge ammunition systems shown loaded	110
2.1.2 (a)	Breech block in the process of closing showing action of beveled face	111
2.1.2 (b)	Breech block in closed position showing cartridge head space (exaggerated).	111
2.1.3	Operation of a typical semi-automatic breech operating mechanism during recoil and counter recoil	112
2.1.4	Crankshaft & crank rotation & breechblock motion	112
2.1.5	Breech locking action of crank roller and sloped portion of T slot in breech block	113
2.1.6 (a)	Simple threaded breech screw	114
2.1.6 (b)	Tapered conical breech screw	114
2.1.6 (c)	Welin breech screw	114
2.1.7	Essential components of breech screw operating mechanism shown exploded	116

2.1.8	Transition of breech screw rotation to swing free motion by cam action	117
2.1.9	Action of the closing spring assembly	118
2.1.10	Action of spring percussion type firing mechanism during cocking and firing	119
2.1.11	Sequence of operation of extractor mechanism	120

2.2

2.2.1	Bending of a thick symmetrically loaded circular plate	122
2.2.2	Axial section of thick circular plate	122
2.2.3	Moments & forces on element of circular plate	124
2.2.4		

2.3

| 2.3.1 | Stresses in interrupted thread breech screw | 130 |

2.5

| 2.5.1 | Components of semi automatic breech operating mechanism | 133 |
| 2.5.2 | Crank and closing spring lever rotation as a result of crankshaft rotation due to operating cam-follower action | 134 |

2.6

2.6.1	Diagram showing geometry of straight edged cam operation at initial contact of follower with cam and after crankshaft rotation by an angle θ	135
2.6.2	Position of crankshaft axis at initial contact between follower & breech operating cam	137
2.6.3	Displacement of crankshaft axis during Phase 1	137
2.6.4	Displacement of crankshaft axis during Phase 2	138
2.6.5	Rotation of crankshaft and related displacement of breech block	140
2.6.6	Relation between crankshaft angle & spring displacement	141

2.7

| 2.7.1 | Torque versus angular displacement curves for 76 mm semi automatic breech mechanism | 146 |

3.1

3.1.1	Hydro-spring recoil system	149
3.1.2	Independent hydro-pneumatic recoil system	150
3.1.3	Dependant hydro-pneumatic recoil system	151

3.2

3.2.1	Graphical representation of recoil forces against recoil displacement	163
3.2.2	Recoil travel-recoil brake orifice area relation	165

3.3

3.3.1	Propellant gas force-time data from shot start to exit	167
3.3.2	Configuration of proposed recoil system	168
3.3.3	Propellant gas force-time curve from shot start to end of gas action	170
3.3.4	Time-propellant gas force data from shot start to end of gas action	171
3.3.5	Worksheet to compute recoil velocity, net braking force and recoil displacement: Worksheet No 1	172
3.3.6	Worksheet for recoil computation: Worksheet No 5	174
3.3.7	Plot of recoil forces versus recoil displacement	179
3.3.8	Tabulated values of displacement, velocity of recoiling parts and net hydraulic braking force	180
3.3.9	Tabulated recoil travel-orifice area data	182

3.4

3.4.1	Counter recoil travel-velocity curves with and without throttling	185

3.5

3.5.1	Values of counter recoil displacement, recoil displacement, recoil velocity and net hydraulic braking force during recoil	189
3.5.2	Schematic layout of the recoil system during counter recoil	190
3.5.3	Construction of workbook for counter recoil computation	190
3.5.4	Determination of counter recoil velocity without throttling and neglecting seal friction force: Worksheet No 1	192
3.5.5	Worksheet to compute throttling force for calculation of seal friction forces	193

3.5.6	Worksheet to calculate seal friction forces	194
3.5.7	Worksheet to compute counter recoil velocity with throttling and seal friction force included	195
3.5.8	Worksheet to compute counter recoil velocity and time with increased throttling	196
3.5.9	Graphical depiction of counter recoil travel versus velocity	197

3.6

3.6.1	Buffer with constant orifice, variable retardation and variable decelerating force	199
3.6.2	Variable orifice area, constant retardation and constant retarding force buffer	200
3.6.3	Counter recoil travel-buffer orifice area relation; variable orifice with constant net retarding force	202
3.6.4	Counter recoil travel-velocity curves for buffing with constant orifice, varying net retarding force and variable orifice-constant net retarding force during buffing	203

3.7

3.7.1	Worksheet for computation of counter recoil velocity during buffing phase; Constant orifice are with variable deceleration	205
3.7.2	Worksheet to compute seal friction forces during buffing	207
3.7.3	Recoil system forces during buffing phase of counter recoil with constant orifice	208
3.7.4	Worksheet for computation of counter recoil velocity and orifice area; variable buffer orifice with constant deceleration	210
3.7.5	Travel-velocity relation for the entire recoil cycle	212

4

4.1

4.1.1	Balancing gear geometry at position of maximum angle of depression	223
4.1.2	Spring force deflection relation	225

4.1.3	$\dfrac{b}{c}$ ratio-maximum energy relation of a balancing gear	228
4.1.4	Worksheet to determine initial spring force and spring stiffness factor	235
4.1.5	Trendline feature in Excel for determination of Spring stiffness Factor and assembled spring force	236

4.2

4.2.1	Sectionized view of a typical pneumatic balancing gear	237
4.2.2	Out of balance and balancing moment for a typical pneumatic balancing gear	240
4.2.3	Friction force in the balancing gear during elevation due to friction torque at the bottom carriage pivot	243
4.2.4	Friction force in the balancing gear during elevation due to friction torque at the cradle bearing	244
4.2.5	Friction force in the balancing gear during depression due to friction torque at the bottom carriage bearing	245
4.2.6	Friction force in the balancing gear during depression due to friction torque at the cradle bearing	246
4.2.7	Plot of out of balance and balancing moments against elevation range for a typical hydro-pneumatic balancing gear	250

4.3

4.3.1	Worksheet to compute torque required to elevate without effect of friction	252
4.3.2	Graphical representation of angle of elevation-out of balance & balancing moment without the effect of friction for the problem stated in Design Exercise 4.3.1	253
4.3.3	Worksheet for calculation of torque including friction effect for pneumatic balancing gear during elevation	255
4.3.4	Worksheet for calculation of torque including friction effect for pneumatic balancing gear during depression	256
4.3.5	Balancing gear moments during manual elevation and depression including effect of friction	257
4.3.6	Worksheet to compute balancing moment and power elevating torque for a hydro-pneumatic balancing gear	259
4.3.7	Worksheet to compute the power torque for depression including the effect of friction	262

4.5

4.5.1	Configuration and dimensions of proposed balancing gear	268
4.5.2	Worksheet to determine spring displacement, Spring Stiffness Factor & assembled spring force	269
4.5.3	Assembled spring force and Spring Stiffness Factor by Trendline feature of Excel	269
4.5.4	Ultimate shear stress of common spring alloys	271
4.5.5	Worksheet to compute torque required for elevation: Spring type balancing gear	272
4.5.6	Worksheet to compute torque required for depression: Spring type balancing gear	273
4.5.7	Plot of put of balance & balancing moments against angle of elevation for the spring type balancing gear of Design Exercise 4.5.1	274

5

5.2

5.2.1	Constant stress cantilever	282
5.2.2	Comparison of bending stress in a gear tooth with a constant stress parabola	282
5.2.3	Lewis Form Factor Y for spur gears	284
5.2.4	Lewis form Factor for worm gears	284
5.2.5	Basic spur gear tooth dimensions in the SI system	286
5.2.6	Wear Factors for spur and bevel gears based on BHN; 20° pressure angle	287
5.2.7	Worm pair Wear Factors for different common materials and lead angle	287
5.2.8	Worm and worm gear force diagram; worm driving in direction as indicated	290
5.2.9	Forces on worm gear tooth with worm driving	291

5.3

5.3.1	Worm based manual elevating mechanism (pitch cylinders depicted)	296
5.3.2	Selection of optimum module based on maximum wear load, bending load and maximum tooth load	299

5.4

5.4.1	Position of gears and pinions in proposed gear train design (pitch cylinders depicted)	302
5.4.2	Worksheet to compute tooth face width of pinion No 1	305
5.4.3	Worksheet to compute tooth face width of pinion No 3	306
5.4.4	Worksheet to compute tooth face width of pinion No 5	307
5.4.5	Worksheet to compute tooth face width of pinion No 7	308
5.4.6	Worksheet to compute torque necessary to accelerate the gear train	309

6

6.1

6.1.1	Pivot type traverse	312
6.1.2	Nut and screw traversing mechanism	313
6.1.3	Pinion & arc traversing mechanism	314
6.1.4	Turret ring traverse	315
6.1.5	Line diagram of a turret traversing mechanism	316

6.2

6.2.1	Bevel gear nomenclature	320

6.3

6.3.1	Worksheet for computation of tooth loads and selection of optimum standard module for worm pair	327
6.3.2	Worksheet for selection of module for bevel gears for traversing gear train	331
6.3.3	Worksheet for selection of module for bevel gears with increased gear ratio	332

6.4

6.4.1	Schematic of gear train for power traversing mechanism	335
6.4.2	Worksheet for selection of module for traversing ring and pinion 1	336
6.4.3	Worksheet for selection of module for gear No 2 and pinion No 3	337
6.4.4	Worksheet for selection of module for gear No 4 and pinion No 5	338
6.4.5	Worksheet for selection of module for gear No 6 and pinion No 7	339
6.4.6	Worksheet for calculation of torque required to accelerate the traversing gear train	340

Contents of CD ROM

af stress.txt		Programme to compute wall thickness, ESP, SF against barrel length for af barrel
autofrettage.txt		Programme to compute stresses in af barrel
CHAMBER.XLS	formulae	Worksheet for chamber design of 76mm gun of (data & formulae)
	results	Worksheet for chamber design of 76mm gun of (data & results)
monobloc.txt		Programme to compute wall thickness, ESP, SF in a monobloc barrel
monostress.txt		Programme to compute stresses in monobloc barrel
pimp.txt		Barrel length-PIMP,CMP curves
PIMP.WK1		76 mm Gun design data sheet in 1-2-3 format
PIMP.XLS	CMP	Interior ballistics: Shot travel-CMP data
	PIMP	76 mm Gun design data sheet
RIFLE.WK1		Internal ballistics worksheet for computation of rifling torque in 1-2-3 format
rifling torque.txt		Programme to compute rifling torque
RIFLING TORQUE.XLS		Internal ballistics worksheet for computation of rifling torque
VIBRATION.XLS	static	Worksheet to determine circular speed: Static loading
	1st	Worksheet to determine circular speed: 1st iteration dynamic loading
	2nd	Worksheet to determine circular speed: 2nd iteration dynamic loading
	3rd	Worksheet to determine circular speed: 3rd iteration dynamic loading
	4th	Worksheet to determine circular speed: 4th iteration dynamic loading

2

breechscrew.txt — Programme to compute stresses: interrupted thread breech screw

crankshaft torque.txt — Programme to compute net clockwise torque in crankshaft

3

braking force.txt — Programme to compute net hydraulic braking force

BUFFER.XLS

    buffer — Worksheet for computation of:
1. Counter recoil velocity during buffing phase; Constant orifice area with variable deceleration
2. Worksheet to compute seal friction forces during buffing

COUNTERRECOIL.XLS

    const — Worksheet for computation of counter recoil velocity and orifice area; variable buffer orifice with constant retardation

    wo throttling — Worksheet for determination of counter recoil velocity without throttling and neglecting seal friction force

    .5%twosealfr — Worksheet to compute:
1. Throttling force for calculation of seal friction forces
2. Seal friction forces

    with seal fr.5%t — Worksheet to compute counter recoil velocity with throttling and seal friction force included

    with.25%t — Worksheet to compute counter recoil velocity and time with increased throttling

orifice.txt — Programme to compute recoil brake orifice area

recoilcalculations.txt		Programme to compute gas force-time relation
	Approx	Worksheet to compute recoil velocity, approximate net braking force and recoil displacement
	1st	Worksheet to compute recoil velocity, precise net braking force and recoil displacement: 1st iteration
RECOIL.XLS	2nd	Worksheet to compute recoil velocity, precise net braking force and recoil displacement: 2nd iteration
	3rd	Worksheet to compute recoil velocity, precise net braking force and recoil displacement: 3rd iteration
	4th	Worksheet to compute recoil velocity, precise net braking force and recoil displacement: 4th iteration
RECOIL_X_V.WK1		Recoil displacement-recoil velocity values in 1-2-3 format
RECOILCALCULATION.WK1		Worksheet to calculate propellant gas force-time data from shot start to exit in 1-2-3 format
RECOILCALCULATIONS.XLS	p-t	Worksheet to calculate propellant gas force-time data from shot start to exit
RECOILORIFICE.WK1		Worksheet for computation of orifice area in 1-2-3 format

4

Em_K.txt		Programme to compute Em-K relation
	Elev torq	Worksheet to compute torque required to elevate without effect of friction
	man ele with fr	Worksheet for calculation of torque, including friction effect for pneumatic balancing gear during elevation
PNEUMATIC.XLS	man dep with fr	Worksheet to calculate torque required for manual depression including the effect of friction
	power ele	Worksheet to compute balancing moment and power elevating torque for a pneumatic balancing gear
	power dep	Worksheet to compute the power torque for depression including the effect of friction
SPRINGFORCE.XLS	stiffness	Worksheet to determine initial spring force and spring stiffness factor

SPRING.XLS	s,P0	Worksheet to determine: 1. spring displacement, Spring Stiffness Factor & assembled spring force
		2. Assembled spring force and Spring Stiffness factor by Trendline feature of Excel
	spring ele	Worksheet to compute torque required for elevation: Spring type balancing gear
	spring dep	Worksheet to compute torque required for depression: Spring type balancing gear

5

WORM.XLS	module	Worksheet for selection of optimum module based on maximum wear load, bending load and maximum tooth load
	e to 1	Worksheet to compute tooth face width of pinion No 1
	2 to 3	Worksheet to compute tooth face width of pinion No 3
POWER ELEVATING GEAR.XLS	4 to 5	Worksheet to compute tooth face width of pinion No 5
	6 to 7	Worksheet to compute tooth face width of pinion No 7
	torque	Worksheet to compute torque necessary to accelerate the gear train

	mod for worm	Worksheet for computation of tooth loads and selection of optimum standard module for worm pair
MANUAL TRAVERSING.XLS	GR 5	Worksheet for selection of module for bevel gears for traversing gear train
	GR 10	Worksheet for selection of module for bevel gears with increased gear ratio
	Tarc to pinion 1	Worksheet for selection of module for traversing ring and pinion 1
	Gear 2 & Pinion 3	Worksheet for selection of module for gear No 2 and pinion No 3
POWER TRAVERSE.XLS	Gear 4 & Pinion 5	Worksheet for selection of module for gear No 4 and pinion No 5
	Gear 6 & Pinion 7	Worksheet for selection of module for gear No 6 and pinion No 7
	Torque	Worksheet for calculation of torque required to accelerate the traversing gear train

xvii

1

Design of Gun Barrels

1.1 General Considerations for Gun Barrel Design

Calibre-wise Classification of Gun Barrels

A generally accepted classification of gun barrels places barrels of calibre above 30 mm in the large caliber or heavy armament category. This category includes tank main guns, artillery guns, howitzers and heavier mortars. Barrels of calibres less than 30 mm fall in the small arms and cannon bracket with hand guns at the lower extremity and anti aircraft cannon at the upper. All other single shot and automatic weapons of intermediate calibre from around 5 mm to 30 mm fall in between.

Regardless of calibre, all gun barrels have similar fundamental features, but, based on their tactical role, manifested in the magnitudes of their muzzle velocities, rates of fire and methods of stabilization of the projectiles involved, barrels of different weapons call for different design and manufacturing approaches.

Effect of Barrel Wear on Accuracy of Guns

Gun barrels must be accurate enough to fulfill their tactical purpose. The accuracy of engagement of a target depends, from the gun aspect, on the velocity with which the projectile exits the barrel. In the case of spin stabilized projectiles, the spin rate, the attitude of the projectile and the forces acting on it at the instant of exit play a significant role towards the realization of this accuracy. Variation in any of these influences results in inaccuracy at the target end.

Erosion of a gun barrel, over a period of time, results in cumulative negative effects of all three factors mentioned above, ultimately rendering the gun ineffectual from the accuracy point of view. Erosion is basically dependant on the temperature attained in the barrel and cannot be eradicated altogether although the effect can be minimized. High temperatures result from a high rate of fire or from the use of high specific energy propellants or a combination of both. Wearing away of the bore surface is caused due to the influence of the hot gases impinging on the bore surface followed by the mechanically abrasive effect of the gases and driving band both moving at high velocities. The effect of erosion is considered at the design stage of the propellant-barrel combination by avoiding abrupt angular surfaces in the path of the gases, cooling additives in the propellant itself, by the alloying of suitable hard elements, such as chromium and nickel to the barrel material, and also by plating the bore surface with abrasion resistant and heat reflecting metals.

Major Influences in the Design of Gun Barrels

The major influences in the design of gun barrels are firstly the internal gas pressure due to combustion of the propellant and secondly, the mechanical properties of the material proposed to be used in its construction. In the design of tank and artillery gun barrels, the propellant gas pressure and the yield strength of the steel are the central deciding factors around which the design revolves.

In high rate of fire weapons, such as machine guns and anti aircraft cannon, thermal stresses due to high temperatures attained as a result of sustained fire, are also taken into consideration.

Safety Factor in Barrel Design

The choice of an optimum safety factor in gun barrel design is of utmost importance.

If the safety factor is low, the working stresses will be too high and the gun barrel will prove weak in service. If the safety factor is too high, the material of the barrel will not be economically stressed at the expense of weight, mobility and cost.

Assuming that the yield point of the material is the basis for determining the safety factor, then the accuracy of assigning a value to the safety factor depends upon the homogeneity of the barrel steel which itself will cause variation in the yield of the steel, the accuracy of determination of the propellant gas pressure, the range over which this maximum gas pressure varies under different ambient conditions and unknown loads like those due to the swaging action of the projectile.

Stresses due to gas pressure being lower towards the muzzle end than those towards the breech end, greater barrel wall thickness than that dictated by the gas pressure alone are necessary. Hence a larger safety factor is accepted for reasons of structural rigidity and strength of the barrel necessary to take the load of muzzle attachments towards the muzzle end.

Stress Effect of Recoil Forces on Gun Barrels

Recoil forces in weapons with single recoil systems do not induce axial stress in excess of the axial stress caused by propellant gas pressure, hence axial stresses due to recoil forces, are not taken into account.

Vibrations of Gun Barrels

During firing, gun barrels are subject to impulse loading which results in vibrations. Propellant gas pressure causes dilative vibrations of the barrel and could result in rupture of the barrel. Driving band pressure on the bore has the same effect on the barrel as the propellant gas pressure.

Barrel or barrel whip is the tendency of the propellant gas pressure and the moving projectile to straighten out the droop due to the intrinsic weight of the barrel. This results in vibrations of the barrel about its intended axis.

These vibrations have two distinct adverse effects. The first is the effect on the material of the barrel, which if in excess, results in structural failure. The second effect is on the accuracy of the weapon.

From the structural point of view, the need for low weight renders the barrel prone to adverse effects of vibration. In the case of artillery and tank gun barrels however, the very bulk of the barrel usually suffices to absorb the stresses caused by the vibrations. In the case of lighter weapons, greater care must be taken during the design stage to create a barrel in which the vibrations are minimized by virtue of its structure.

In low rates of fire guns such as artillery and tank main armaments, the rate of fire is limited by constraints such as the speed of loading and the need to re-aim the weapon. Subsequently, the vibrations caused by the firing of a round damp out naturally before the firing of the following round. The effect on accuracy in the case of rapid fire weapons is pronounced and due care is taken at the design stage to minimize the undesirable effect of barrel vibrations on accuracy.

Heating of Gun Barrels

Combustion of the propellant in the chamber, results in the release of a huge quantity of heat inside the bore. Some of this heat is usefully utilized towards propulsion of the projectile whereas the major part is carried away partly by the exhaust gases and partly by conduction to the walls of the barrel.

The factors which affect heat transfer to the material of the gun barrel are the propellant characteristics, the thickness of the barrel wall, its thermal conductivity and the rate and continuity of fire.

The heat transferred to the gun barrel affects the material of the gun barrel in two ways. Uniform heating of the barrel wall decreases the yield strength of the material whereas differential heating introduces stresses in the material.

Continuous heating of the bore can result in increased clearance between the driving band and the bore surface, a condition called barrel dilation, which leads to leakage of gas past the projectile and oscillations of the projectile about its longer axis during its travel up the bore, ultimately resulting in a loss of accuracy.

Bending of Gun Barrels

Differential heating or cooling of the barrel due to environmental influences such as solar heat and rain or snow cause the barrel to bend away from its intended axis and reduce accuracy. Barrel bend cannot be predicted at the design stage, for obvious reasons. It can however be minimized by thermal jacketing of the barrel with insulating material and corrected for by the use of muzzle reference devices incorporated in the fire control system especially in the case of high accuracy weapons.

Cook Off

Cook off in gun barrels is the spontaneous ignition of the propellant, explosive components of the fuze, or the shell main filling itself, due to heat transfer from the hot barrel wall, the barrel wall having attained a high temperature as a result of continuous firing. Cook off may be prevented at the design stage of the ammunition by fixing the cook off temperature of the explosive components of the ammunition. Risk from cook off is minimized during firing by barrel cooling which may vary from simply swabbing the chamber with a wet mop, to air or water-cooling as in the case of automobile engines.

1.2 Theories of Failure of Gun Barrel Materials

Basis for Failure Theories

All mechanical components are subject to complex stresses during the usage for which they are designed. Materials of which such components are comprised are investigated for failure by simplified testing methods. For instance, tensile testing of ductile materials, compressive testing of brittle materials and limited tests related to the strength of materials in shear, all of which are performed on standard specimens. These tests do not take into account conditions of complex stressing of the material. Strength theories aim to relate the failure of materials under conditions of combined stress to their failure characteristics determined during simple tension or compression testing.

Whenever a body is subject to complex stresses, it is possible to resolve the stress into three principal stresses in three mutually orthogonal directions, as in Fig 1.2.1. The stress condition of the body can be defined by the magnitude of the three principal stresses $\sigma_x, \sigma_y, \sigma_z$ as depicted in the figure below. Algebraically it is also assumed that $\sigma_x \rangle \sigma_y \rangle \sigma_z$

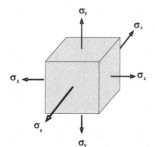

Fig 1.2.1 Resolution of combined stresses into three principal directions

Theories of Failure Associated with Gun Barrel Materials.

Two theories that afford acceptable agreement with experiment in the case of ductile materials like gun barrel steel are investigated here. They are the Maximum Shear Stress or Tresca Theory and the Maximum Distortion Energy or the Huber-Von Mises-Hencky Theory.

The Maximum Shear Stress or Tresca Theory

This theory is based on the assumption that the elastic failure of the material occurs when the greatest shear stress reaches the magnitude of maximum shear stress at yield point in a simple tensile test.

The maximum shear stress being equal to half the difference between the maximum and minimum principal stresses, and since the maximum shear stress in a simple tensile test is equal to half the normal stress, the condition for failure according to this theory becomes:

$$\frac{1}{2}\left(\sigma_{max} - \sigma_{min}\right) = \tau_{max} = \frac{1}{2}\sigma_{yp} \dotfill [1.2.1]$$

σ_{yp} : yield stress of the material in simple tension

τ_{max} : maximum shear stress

The natures of the stresses, whether compressive or tensile, are taken into account and the maximum stress difference is equated to the maximum shear stress. In the case of a two-dimensional stress system the problem becomes quite straight forward. This theory has been found to agree well with experimental result.

The Maximum Distortion Energy or Huber-Von Mises-Hencky Theory.

Failure, according to this theory, is predicted to occur when the internal distortion energy reaches the distortion energy at the yield point in a simple tensile test.

In the distortion energy criterion, the total strain energy is split into two parts; one related to the change in volume and the second to distortion. It is the second part only which is considered.

$$U = U_v + U_\tau \dotfill [1.2.2]$$

U: net strain energy of the system
U_v: volumetric strain energy
U_τ : energy of distortion

Consider three mutually perpendicular stresses $\sigma_x, \sigma_y, \sigma_z$ acting on a cube as shown in Figure 1.2.1:

The total strain produced in the direction of x is:

$$\varepsilon_x = \frac{\sigma_x}{E} - \frac{\mu[\sigma_y + \sigma_z]}{E}$$

μ : Poissons ratio
E : Youngs Modulus.

The strain produced in the direction of y is:

$$\varepsilon_y = \frac{\sigma_y}{E} - \frac{\mu[\sigma_x + \sigma_z]}{E}$$

The strain in the direction of z is:

$$\varepsilon_z = \frac{\sigma_z}{E} - \frac{\mu[\sigma_x + \sigma_y]}{E}$$

The volumetric strain becomes:

$$\varepsilon_v = \varepsilon_x + \varepsilon_y + \varepsilon_z$$

Or:

$$\varepsilon_v = \frac{\sigma_x + \sigma_y + \sigma_z}{E}(1 - 2\mu) \quad \text{..} \quad [1.2.3]$$

With reference to Fig 1.2.2 ahead under a gradually applied tensile load, the stress-strain curve is a straight line up to the limit of proportionality. Hence the strain energy is equal to the area, shown shaded under the stress-strain curve. It follows that:

8

Strain energy = $\dfrac{1}{2}$ stress.strain

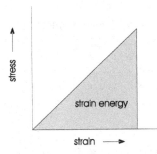

Fig 1.2.2: Strain energy within the limit of proportionality

The strain energy per unit volume due to stress σ_x is:

$$U_x = \frac{1}{2}\sigma_x \varepsilon_x$$

Or:

$$U_x = \frac{\sigma_x^2 - \sigma_x \mu[\sigma_y + \sigma_z]}{2E}$$

Similar expressions hold good for the strain energy due to σ_y, σ_z. Hence:

$$U_y = \frac{\sigma_y^2 - \sigma_y \mu[\sigma_x + \sigma_z]}{2E}$$

$$U_z = \frac{\sigma_z^2 - \sigma_z \mu[\sigma_x + \sigma_y]}{2E}$$

The net strain energy per unit volume is therefore:

$$U = \frac{\sigma_x^2 + \sigma_y^2 + \sigma_z^2 - 2\mu[\sigma_x\sigma_y + \sigma_y\sigma_z + \sigma_z\sigma_x]}{2E} \quad\dotfill\quad [1.2.4]$$

The principal stresses of the system depicted in Fig 1.2.1 can be split into two systems of principal stresses:

$(\sigma_x - \sigma), (\sigma_y - \sigma), (\sigma_z - \sigma)$ and σ, σ, σ superimposed on each other as depicted in Fig 1.2.3 by the introduction of a fourth stress denoted by σ, where:

$$\sigma = \frac{1}{3}\left(\sigma_x + \sigma_y + \sigma_z\right)$$

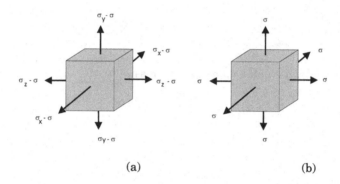

(a)   (b)

Fig 1.2.3: Superimposition of a fourth stress σ on an existing stress system

From Equation [1.2.3], volumetric strain of the system in Fig 1.2.1:

$$\varepsilon_v = \frac{\sigma_x + \sigma_y + \sigma_z}{E}\left(1 - 2\mu\right)$$

Volumetric strain of the system in Fig 1.2.3(a) is found by replacing $\sigma_x, \sigma_y, \sigma_z$ by: $\sigma_x - \sigma, \sigma_y - \sigma, \sigma_z - \sigma$ respectively:

$$\varepsilon_a = \frac{1}{E}\left(1 - 2\mu\right)\left[\left(\sigma_x - \sigma\right) + \left(\sigma_y - \sigma\right) + \left(\sigma_z - \sigma\right)\right]$$

$$= \frac{1 - 2\mu}{E}\left(\sigma_x + \sigma_y + \sigma_z - 3\sigma\right) = 0$$

Similarly volumetric strain of the system in Fig 1.2.3(b) is:

$$\varepsilon_b = \frac{3\sigma}{E}(1-2\mu) \ldots\ldots\ldots\ldots\ldots\ldots\ldots\ldots\ldots\ldots\ldots\ldots\ldots\ldots\ldots\ldots\ldots \text{[1.2.5]}$$

It follows that the volumetric strain of the system is identical to the volumetric strain of the system in Fig 1.2.3(b). It is also evident that there is no shear strain in the system depicted in Fig 1.2.3(b) due to the fact that the principal stresses are equal.

Evidently the distortion strain energy of the system per unit volume is the net strain energy of the system per unit volume minus the strain energy of the system, per unit volume, where the principal stresses are equal i.e. as in Fig 1.2.3(b).

Applying Equation [1.2.4] and substituting $\sigma_x = \sigma_y = \sigma_z = \sigma$, in the special case of Fig 1.2.3(b), volumetric strain energy per unit volume is:

$$U_v = \frac{3\sigma^2 - 6\mu\sigma^2}{2E} = \frac{3\sigma^2}{2E}(1-2\mu)$$

Or:

$$U_v = \frac{(1-2\mu)}{6E}(\sigma_x + \sigma_y + \sigma_z)^2 \ldots\ldots\ldots\ldots\ldots\ldots\ldots\ldots\ldots\ldots\ldots\ldots\ldots\ldots \text{[1.2.6]}$$

Therefore from Equation [1.2.2], the distortion energy per unit volume is:

$$U_\tau = U - U_v$$

Or:

$$U_\tau = \frac{\sigma_x^2 + \sigma_y^2 + \sigma_z^2 - 2\mu[\sigma_x\sigma_y + \sigma_y\sigma_z + \sigma_z\sigma_x]}{2E} - \frac{(1-2\mu)}{6E}(\sigma_x + \sigma_y + \sigma_z)^2$$

$$= \frac{\sigma_x^2 + \sigma_y^2 + \sigma_z^2 - 2\mu[\sigma_x\sigma_y + \sigma_y\sigma_z + \sigma_z\sigma_x]}{2E} - \frac{(1-2\mu)}{6E}\left[\sigma_x^2 + \sigma_y^2 + \sigma_z^2 + 2(\sigma_x\sigma_y + \sigma_y\sigma_z + \sigma_x\sigma_z)\right]$$

$$= \frac{1}{6E}\left[\left(\sigma_x^2 + \sigma_y^2 + \sigma_z^2\right)\left(2 + 2\mu\right) - \left(2 + 2\mu\right)\left(\sigma_x\sigma_y + \sigma_y\sigma_z + \sigma_z\sigma_x\right)\right]$$

$$= \frac{1}{6E}\left[\left(\sigma_x^2 + \sigma_y^2 + \sigma_z^2\right)2\left(1 + \mu\right) - 2\left(1 + \mu\right)\left(\sigma_x\sigma_y + \sigma_y\sigma_z + \sigma_z\sigma_x\right)\right]$$

$$= \frac{\left(1 + \mu\right)}{6E}\left[2\left(\sigma_x^2 + \sigma_y^2 + \sigma_z^2\right) - 2\left(\sigma_x\sigma_y + \sigma_y\sigma_z + \sigma_z\sigma_x\right)\right]$$

Or:

$$U_\tau = \frac{1}{12G}\left[\left(\sigma_x - \sigma_y\right)^2 + \left(\sigma_y - \sigma_z\right)^2 + \left(\sigma_z - \sigma_x\right)^2\right] \dots\dots\dots\dots\dots\dots\dots\dots\dots \text{[1.2.7]}$$

$$G = \frac{E}{2\left(1 + \mu\right)} : \text{Bulk Modulus}$$

According to Equation [1.2.7], distortion energy per unit volume of the material is given by:

$$U_\tau = \frac{1}{12G}\left[\left(\sigma_x - \sigma_y\right)^2 + \left(\sigma_y - \sigma_z\right)^2 + \left(\sigma_z - \sigma_x\right)^2\right]$$

Applying this condition to a simple tensile test where only one principal stress exists:

$$U_\tau = \frac{1}{12G}\left[\left(\sigma_{yp} - 0\right)^2 + 0 + \left(0 - \sigma_{yp}\right)^2\right]$$

Or:

$$U_\tau = \frac{\sigma_{yp}^2}{6G} \dots\dots\dots\dots\dots\dots\dots\dots\dots\dots\dots\dots\dots\dots\dots\dots \text{[1.2.8]}$$

Combining Equations [12.7] and [12.8], according to the Maximum Shear Energy or Huber-Von Mises-Hencky Theory:

$$2\sigma_{yp}^2 = \left(\sigma_x - \sigma_y\right)^2 + \left(\sigma_y - \sigma_z\right)^2 + \left(\sigma_z - \sigma_x\right)^2 \dots\dots\dots\dots\dots\dots\dots \text{[1.2.9]}$$

Two Dimensional Stress Case

In the two dimensional stress case, where $\sigma_z = 0$, Equation [1.2.8] simplifies to:

$$\sigma_{yp}^2 = \sigma_x^2 - \sigma_x \sigma_y + \sigma_y^2 \quad \text{..} \quad [1.2.10]$$

This theory is a refinement on the Beltrami Theory and had been found to give the best agreement with experimental results performed on ductile materials.

1.3 Conventions Used In Gun Barrel Design

Gun Pressure Codes

Gun barrel designers have established gun pressure codes in order to standardize the way in which the results of interior ballistics calculations are interpreted and made use of by the gun designer community. Commonly, various pressures used in the course of design of gun barrels for low rate of fire weapons are defined as follows:

Computed Maximum Pressure (CMP)

The computed maximum chamber pressure attained, as a result of internal ballistics computation, at a specified ambient temperature, usually 70° F, in order to achieve the stipulated muzzle velocity.

Rated Maximum Pressure (RMP)

This is the maximum pressure that should not be exceeded by the average of the maximum individual pressures of a group of rounds. It takes into account the variation in pressure from round to round within a group of rounds.

Practically:

RMP = CMP + 16.548 MPa.. [1.3.1]

Permissible Individual Maximum Pressure (PIMP)

This is the maximum pressure, which may be achieved by any single round under the most adverse conditions of employment. This pressure forms the basis for the barrel design and is the effective pressure from the breech end of the chamber to the point of maximum pressure along the bore.

Practically:

PIMP = 1.15(CMP+16.548)

Or:

PIMP = 1.15CMP + 24.822 MPa .... [1.3.2]

Allowable Stress

In practice, a stress lower than that at the yield point is usually laid down. In the case of heavy, slow firing weapons this allowable stress is designated as 69.85 M Pa less than the yield stress at a temperature of 70^0 F and is the limiting stress magnitude for design of the entire barrel.

By definition:

$$\sigma_a = \sigma_{yp} - 68.95 \text{ MPa}$$ .... [1.3.3]

Elastic Strength Pressure (ESP)

This is the actual pressure which the barrel can sustain and is computed from Equation [1.4.5] in the case of monobloc barrels and Equation [1.6.9] in the case of autofrettage, by inserting the value of the final wall ratio.

Safety Factor (SF)

The safety factor at any section is given by:

$$SF = \frac{ESP}{PIMP}$$ .... [1.3.4]

14

Allowance for Eccentricity

Eccentricity between the circle of the bore and the circle of the outer diameter at any cross section results in a reduction of the barrel thickness at a radius and consequent increase in the thickness at the radius diametrically opposite.

Allowance is made for this unavoidable eccentricity by adjusting the minimum wall thickness at any point so that it is not less than 90% of the wall thickness diametrically opposite. In other words:

$0.9t_{adj} \not< t$

Or: $t_{adj} \not< 1.111t$

t: minimum wall thickness from stress considerations
t_{adj}: adjusted barrel wall thickness; b_{adj}: adjusted outer radius

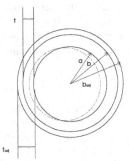

Fig 1.3.1: Eccentricity of bore and outer surface circles of a gun barrel

Or:

$$t_a = \frac{1}{K}t$$

K can assume any value greater than 0.9 and less than 1.0. Practically a value of $K = 0.95$ is chosen. Hence:

$$t_{adj} = 1.0526t$$

Or:

$$b_{adj} - a = 1.0526(b - a)$$

$$T_{adj} = 1.0526T - .0526 \dotfill [1.3.5]$$

b_{adj}: adjusted outer radius

$T_{adj} = \dfrac{b_{adj}}{a}$: wall radius ratio adjusted for eccentricity

Droop, Attachments and Manufacturing Influences on Gun Barrel Wall Thickness

Finally the walls of the gun barrel will invariably be thicker than the minimum computed thickness at all sections except for some distance from the breech end where the barrel is continuously threaded for mating with the breech ring. Buttress threads are ideally suited for the purpose. As the thrust is unidirectional they provide the maximum contact area and afford reduction in stresses.

For ease of manufacturing, curvatures, which occur along the length of the barrel, as a result of the dependency of barrel dimensions on the space pressure curve, are replaced with cylindrical sections or frustums of cones. The barrel is also invariably thickened to minimize droop. Towards the muzzle end, the barrel may be thickened for structural strength in order to sustain the load of muzzle attachments and to make allowance for cutting of the threading necessary for mounting them. Cylindrical surfaces are also provided for attachment of recoil system mounting brackets and for the fitting of slide brackets.

Allowances for manufacturing tolerances are incorporated in the final design dimensions. In practice the ESP and Safety Factor are calculated from the final dimensions of the gun barrel arrived at after all loads and tolerances have been applied.

1.4 Monobloc Gun Barrel Design

Stresses in Gun Barrel Design

Since a gun barrel is an open cylinder, stress in the axial direction due to propellant gas pressure is zero. Some axial stress is induced by the recoil force and also by the friction force between the projectile and the bore surface. However in the ultimate analysis, the magnitude of the axial stress is negligible compared to the magnitude of the hoop and radial stresses and is omitted from the stress analysis.

General Equations for Hoop and Radial Stresses in a Monobloc Barrel

The hoop stress in a monobloc gun barrel, subject to an internal pressure, is given by:

$$\sigma_t = p_i \frac{a^2}{b^2 - a^2}\left(1 + \frac{b^2}{r^2}\right) \dots\dots\dots\dots\dots\dots\dots\dots\dots\dots \text{Equation [1.2.13] Reference 1}$$

The radial stress is given by:

$$\sigma_r = p_i \frac{a^2}{b^2 - a^2}\left(1 - \frac{b^2}{r^2}\right) \dots\dots\dots\dots\dots\dots\dots\dots \text{Equation [1.2.12] Reference 1}$$

p_i: *PIMP*
a: inner radius of the gun barrel
b: outer radius of the gun barrel
r: any intermediate radius between a and b

Maximum Hoop and Radial Stresses in a Monobloc Barrel

In a monobloc barrel the maximum stresses occur at the bore surface, i.e. when $r = a$. The maximum hoop and radial stresses are hence:

$$\sigma_t = p_i \frac{a^2}{b^2 - a^2}\left(1 + \frac{b^2}{a^2}\right)$$

Or:

$$\sigma_t = p_i \frac{b^2 + a^2}{b^2 - a^2} \quad \text{............} \quad [1.4.1]$$

And:

$$\sigma_r = -p_i \quad \text{............} \quad [1.4.2]$$

If the wall ratio or the ratio of outer to inner radius of any cross section of the barrel is:

$$T = \frac{b}{a} \quad \text{............} \quad [1.4.3]$$

Then in terms of the wall ratio T Equation [1.4.1] becomes:

$$\sigma_{t\,max} = p_i \frac{T^2 + 1}{T^2 - 1} \quad \text{............} \quad [1.4.4]$$

Applying the Huber-Von Mises-Hencky criterion which defines the equivalent stress in materials subject to combined loading, in this case by hoop and radial stress, by substituting the expressions for maximum hoop and radial stresses from Equations [1.4.1] and [1.4.2] and the allowable stress of the steel for the equivalent stress, Equation [1.2.10] becomes:

$$\sigma_a^2 = p_i^2 + p_i^2 \left(\frac{T^2 + 1}{T^2 - 1} \right) + p_i^2 \left(\frac{T^2 + 1}{T^2 - 1} \right)^2$$

σ_a : allowable stress of the steel

Or: $\dfrac{\sigma_a^2}{p_i^2} = \dfrac{3T^4 + 1}{\left(T^2 - 1\right)^2}$

From which:

$$\frac{p_i}{\sigma_a} = \frac{T^2 - 1}{\sqrt{3T^4 + 1}} \quad\text{...} \quad [1.4.5]$$

Solving for T:

$$\left(3\frac{p_i^2}{\sigma_a^2} - 1\right)T^4 + 2T^2 + \frac{p_i^2}{\sigma_a^2} - 1 = 0$$

From which:

$$T^2 = \frac{1 + 2\dfrac{p_i}{\sigma_a}\sqrt{1 - \dfrac{3}{4}\left(\dfrac{p_i^2}{\sigma_a^2}\right)}}{1 - 3\dfrac{p_i^2}{\sigma_a^2}}$$

Or:

$$T = \left[\frac{1 + 2\dfrac{p_i}{\sigma_a}\sqrt{1 - \dfrac{3}{4}\left(\dfrac{p_i^2}{\sigma_a^2}\right)}}{1 - 3\dfrac{p_i^2}{\sigma_a^2}}\right]^{\frac{1}{2}} \quad\text{...} \quad [1.4.6]$$

For a given gun barrel, the inner radius to the land surface or calibre and the chamber inner dimensions being specified, the wall ratio thereafter dictates the minimum outer diameter and consequently the minimum barrel thickness. The minimum outer radius based solely on the stress analysis is now determined from Equation [1.4.3]:

1.5 Design of a Monobloc Gun Barrel

Design Exercise 1.5.1

The feasibility study for an anti tank gun, based on the tactical requirement against an established level of armour, culminated in the requirement for a projectile of calibre 76.0 mm, mass 6.2 Kg and of muzzle velocity 680 m/s. Based on the specifications and results of the internal ballistics computations the preliminary design for a monobloc barrel is to be carried out. Necessary internal ballistics information is contained in the Worksheet of Fig 1.5.1.

Symbols used in the worksheet:

mv: muzzle velocity
mc: charge mass
mp: projectile mass
d: calibre
A: bore cross sectional area
p_max: maximum propellant gas pressure
p_bar: pressure ratio
s: instantaneous shot travel
s_max: shot travel at instant of *p_max*
s_mz: shot travel up to muzzle
netap, sigmanetap, psi: ballistic factors

$$lamda = \frac{s}{s_{max}}$$

Interior ballistics: Shot travel vs CMP data			
Data:			
mv	680.00	d	0.0762
mc	1.08	A	4.56E-03
p_max	2.27E+08	mp	6.20
s_max	0.226	p_bar	1.32E+08
s_mz	2.5870	sigmanetap	0.0875
netap	5.82E-01		
lamda	**s**	**psi**	**CMP**
0.00	0.000	0.000	0.00E+00
0.25	0.057	0.741	1.68E+08
0.50	0.113	0.912	2.07E+08
0.75	0.170	0.980	2.22E+08
1.00	0.226	1.000	2.27E+08
1.25	0.283	0.989	2.24E+08
1.50	0.340	0.965	2.19E+08
1.75	0.396	0.932	2.12E+08
2.00	0.453	0.898	2.04E+08
2.50	0.566	0.823	1.87E+08
3.00	0.679	0.747	1.70E+08
3.50	0.792	0.675	1.53E+08
4.00	0.905	0.604	1.37E+08
4.50	1.019	0.546	1.24E+08
5.00	1.132	0.495	1.12E+08
6.00	1.358	0.403	9.15E+07
7.00	1.585	0.338	7.67E+07
8.00	1.811	0.284	6.44E+07
9.00	2.037	0.248	5.63E+07
10.00	2.264	0.220	4.99E+07
11.00	2.490	0.199	4.52E+07
11.43	2.587	0.191	4.34E+07

Fig 1.5.1: Interior ballistics: Shot travel-CMP data

Internal Dimensions of Barrel

The internal ballistics solution for the design of the barrel yielded the theoretical dimensions of chamber and bore given diagrammatically in Fig 1.5.2.

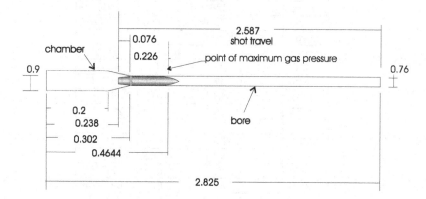

Fig 1.5.2: Inner dimensions of monobloc gun barrel from the internal ballistics solution

Design Procedure

Computation of Gun Design Data

The gun design data is the magnitude of the PIMP at successive sections, of known internal radius, along the length of the barrel, commencing from the breech end. The PIMP is computed by the gun designer from the CMP contained in the interior ballistics worksheet with the help of Equation [1.3.2]. The internal dimensions of the barrel provided in Fig 1.5.2. are entered into the worksheet against corresponding values of shot travel.

Additional symbols used in worksheet:
l: barrel length measured from breech face; d: internal diameter
l0: barrel length to base of projectile before shot start; ri: internal radius

Interior ballistic: Shot travel vs PIMP							
Data:							
mv	680.00	mp	6.20	s_mz	2.5870	p_max	2.27E+08
mc	1.08	l0	0.238	sigmanetap	0.0875	s_max	0.2264
lamda	**s**	**l**	**psi**	**CMP**	**d**	**ri**	**PIMP**
0.00	0.000	0.000	0.000	0.000	0.090	0.045	2.86E+08
0.00	0.000	0.200	0.000	0.000	0.090	0.045	2.86E+08
0.00	0.000	0.238	0.000	2.27E+07	0.083	0.042	2.86E+08
0.25	0.057	0.295	0.741	1.68E+08	0.076	0.038	2.86E+08
0.50	0.113	0.351	0.912	2.07E+08	0.076	0.038	2.86E+08
0.75	0.170	0.408	0.980	2.22E+08	0.076	0.038	2.86E+08
1.00	0.226	0.464	1.000	2.27E+08	0.076	0.038	2.86E+08
1.25	0.283	0.521	0.989	2.24E+08	0.076	0.038	2.83E+08
1.50	0.340	0.578	0.965	2.19E+08	0.076	0.038	2.77E+08
1.75	0.396	0.634	0.932	2.12E+08	0.076	0.038	2.68E+08
2.00	0.453	0.691	0.898	2.04E+08	0.076	0.038	2.59E+08
2.50	0.566	0.804	0.823	1.87E+08	0.076	0.038	2.40E+08
3.00	0.679	0.917	0.747	1.70E+08	0.076	0.038	2.20E+08
3.50	0.792	1.030	0.675	1.53E+08	0.076	0.038	2.01E+08
4.00	0.905	1.143	0.604	1.37E+08	0.076	0.038	1.82E+08
4.50	1.019	1.257	0.546	1.24E+08	0.076	0.038	1.67E+08
5.00	1.132	1.370	0.495	1.12E+08	0.076	0.038	1.54E+08
6.00	1.358	1.596	0.403	9.15E+07	0.076	0.038	1.30E+08
7.00	1.585	1.823	0.338	7.67E+07	0.076	0.038	1.13E+08
8.00	1.811	2.049	0.284	6.44E+07	0.076	0.038	9.89E+07
9.00	2.037	2.275	0.248	5.63E+07	0.076	0.038	8.95E+07
10.00	2.264	2.502	0.220	4.99E+07	0.076	0.038	8.22E+07
11.00	2.490	2.728	0.199	4.52E+07	0.076	0.038	7.68E+07
11.43	2.587	2.825	0.191	4.34E+07	0.076	0.038	7.48E+07

Fig 1.5.3: 76 mm Gun design data sheet

Gun Design Curve

The gun design curve is computed with the help of a computer programme. In order to illustrate the difference between the design curve and the interior ballisticians curve, the CMP curve is also plotted.

Matlab Programme 1.5.1

To run this programme using Matlab, save Worksheet: gun data as a WK1 file to a: drive. Copy and paste the contents of PIMP.txt to the Matlab command window. Press enter. Similarly for other programmes needing data from Excel files.

```
% programme to plot barrel length- pimp/cmp curves
% file name: PIMP.txt

range1='h6..h29'
pimp=wk1read('a:pimp',5,7,range1)
range2='c6..c29'
l=wk1read('a:pimp',5,2,range2)
range3='e6..e29'
pcmp=wk1read('a:pimp',5,4,range3)
plot(l,pimp,l,pcmp)
title('Barrel length vs pressure curve')
xlabel('barrel length');ylabel('pressure')
gtext('computed maximum pressure');gtext('permissible individual maximum
pressure')
```

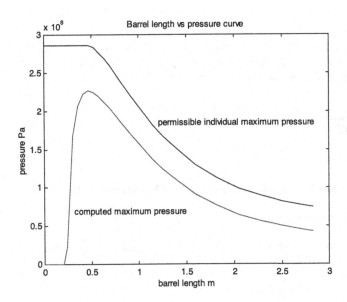

Fig 1.5.4: Barrel length-CMP, PIMP curves

Computation of Wall Thickness

The yield stress of the steel used, AISI 4150 tempered at 650 C is 841.0 M Pa. This is reduced to an allowable stress of 772.05 MPa, according to the definition of allowable stress.

Wall thickness ratio is computed from the gun design data with the help of Equation [1.4.6].

The outer radius of the barrel is computed with the help of Equation [1.4.3].

ESP

ESP is computed with the help of Equation [1.4.5] inserting the value of the final wall ratio after making allowance for eccentricity during manufacture by applying Equation [1.3.5] as given below:

$$ESP = \sigma_a \frac{T_{adj}^2 - 1}{\sqrt{3T^4 + 1}}$$

Safety Factor

The Safety Factor is computed with the help of Equation [1.3.4].

Matlab Programme 1.5.2

```
% programme to compute wall thickness, ESP, SF in a monobloc barrel
% file name: monobloc.txt

sigma_a=772.05.*10.^6 % Pa
range1='h6..h29'
p_i=wk1read('a:pimp',5,7,range1)
range2='c6..c29'
l=wk1read('a:pimp',5,2,range2)
range3='g6..g29'
r_i=wk1read('a:pimp',5,6,range3)
P=p_i./sigma_a
PP=P.^2
T=sqrt((1+2.*P.*sqrt(1-.75.*PP))./(1-3.*PP))
T_adj=1.0526.*T-.0526
r_o=T_adj.*r_i
ESP=sigma_yp.*(T_adj.^2-1)./sqrt(3.*T_adj.^4+1)
SF=ESP./p_i
subplot('position',[.1,.4,.3,.4]),plot(l,ESP,'k',l,p_i,'k');title('          (a) barrel length
vs ESP & PIMP');xlabel('barrel length m');ylabel('ESP Pa')
subplot('position',[.6,.4,.3,.4]), plot(l,SF);title('    (b) barrel length vs safety
factor');xlabel('barrel length m');ylabel('safety factor')
```

subplot('position',[.1,.1,.8,.15]),plot(l,r_i,l,r_o);title('(c) barrel length vs wall thickness');xlabel('barrel length m');ylabel('wall thickness m')

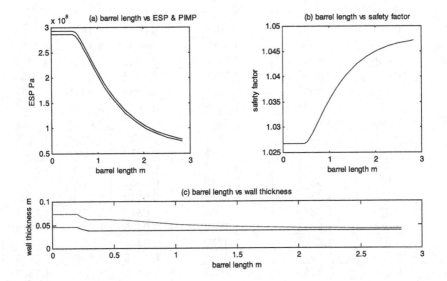

Fig 1.5.5: Barrel length-ESP, PIMP, Safety Factor, wall thickness curves

Barrel length m	Inner radius m	Wall thickness ratio	Adjusted wall thickness ratio	Outer radius m	ESP x10⁸	Safety Factor
0.000	0.045	1.6997	1.7365	0.0781	3.1874	1.1153
0.200	0.045	1.6997	1.7365	0.0781	3.1874	1.1153
0.238	0.042	1.6997	1.7365	0.0721	3.1874	1.1153
0.295	0.038	1.6997	1.7365	0.0661	3.1874	1.1153
0.351	0.038	1.6997	1.7365	0.0662	3.1874	1.1153
0.408	0.038	1.6997	1.7365	0.0662	3.1874	1.1153
0.464	0.038	1.6997	1.7365	0.0660	3.1874	1.1153
0.521	0.038	1.6848	1.7209	0.0654	3.1565	1.1157
0.578	0.038	1.6538	1.6881	0.0641	3.0890	1.1165
0.634	0.038	1.6137	1.6459	0.0625	2.9959	1.1177
0.691	0.038	1.5752	1.6055	0.0610	2.8999	1.1189
0.804	0.038	1.4991	1.5253	0.0580	2.6870	1.1214
0.917	0.038	1.4318	1.4545	0.0553	2.4702	1.1240
1.030	0.038	1.3754	1.3952	0.0530	2.2636	1.1263
1.143	0.038	1.3255	1.3426	0.0510	2.0590	1.1285
1.257	0.038	1.2882	1.3034	0.0495	1.8911	1.1303
1.370	0.038	1.2577	1.2712	0.0483	1.7430	1.1318
1.596	0.038	1.2071	1.2180	0.0463	1.4747	1.1345
1.823	0.038	1.1744	1.1835	0.0450	1.2843	1.1363
2.049	0.038	1.1488	1.1566	0.0440	1.1256	1.1377
2.275	0.038	1.1325	1.1395	0.0433	1.0196	1.1387
2.502	0.038	1.1202	1.1265	0.0428	0.9370	1.1394
2.728	0.038	1.1112	1.1170	0.0424	0.8749	1.1399
2.825	0.038	1.1080	1.1136	0.0423	0.8526	1.1401

Fig 1.5.6: Tabulated design data of monobloc gun barrel

Design Exercise 1.5.2

Stress Analysis of Monobloc Gun Barrel

Based on the obtained outer radius at various sections along the barrel, the magnitude of the tensile, radial and equivalent stresses is verified applying the design pressure values at the respective sections and making use of the following equations:

Tensile stress:

$$\sigma_t = p_i \frac{a^2}{b^2 - a^2}\left(1 + \frac{b^2}{r^2}\right) \dots\dots\dots\dots\dots\dots\dots\dots\dots\dots\dots\dots\dots \text{Equation [1.2.13] Reference 1}$$

Radial stress:

$$\sigma_r = p_i \frac{a^2}{b^2 - a^2}\left(1 - \frac{b^2}{r^2}\right) \dots\dots\dots\dots\dots\dots\dots\dots\dots\dots\dots \text{Equation [1.2.12] Reference 1}$$

For the equivalent stress, Equation [1.2.9] is used suitably modified:

$$\sigma_{eq}^2 = \sigma_t^2 - \sigma_r\sigma_t + \sigma_r^2$$

Computation of Stresses

Three representative sections of the barrel are considered. The first is selected at the centre of the chamber which is under the maximum pressure, the second at the mid section of the barrel and the third at the muzzle end. The stresses are computed using a computer programme.

First section at 0.464 m from breech end

Inner radius: 0.038 m
Outer radius: 0.0654 m
p_i: 286 M Pa

Second section at 1.5962 m from breech end

Inner radius: 0.038 m
Outer radius: 0.0463 m
p_i: 130 M Pa

Third section at muzzle end

Inner radius: 0.038 m
Outer radius: 0.0423 m
p_i: 74.8 M Pa

Matlab Programme 1.5.3

```
%programme to compute stresses in monobloc barrel
% filename: monostress.txt
sigmayp=.841.*10.^9
% at p max
a1=.038;b1=.0615
p1=286.*10^6
r1=linspace(a1,b1,100)
sigmat1=a1.^2.*p1/(b1.^2-a1.^2).*(1+(b1.^2./r1.^2))
sigmar1=a1.^2.*p1/(b1.^2-a1.^2).*(1-(b1.^2./r1.^2))
sigmaeq1=sqrt(sigmat1.^2-sigmat1.*sigmar1+sigmar1.^2)

% at mid section
a2=.038;b2=.0455
p2=130.*10.^6
r2=linspace(a2,b2,100)
sigmat2=a2.^2.*p2/(b2.^2-a2.^2).*(1+(b2.^2./r2.^2))
sigmar2=a2.^2.*p2/(b2.^2-a2.^2).*(1-(b2.^2./r2.^2))
sigmaeq2=sqrt(sigmat2.^2-sigmat2.*sigmar2+sigmar2.^2)

% at muzzle
a3=.038;b3=.0419
p3=74.8.*10.^6
r3=linspace(a3,b3,100)
sigmat3=a3.^2.*p3/(b3.^2-a3.^2).*(1+(b3.^2./r3.^2))
```

sigmar3=a3.^2.*p3/(b3.^2-a3.^2).*(1-(b3.^2./r3.^2))
sigmaeq3=sqrt(sigmat3.^2-sigmat3.*sigmar3+sigmar3.^2)

subplot('position',[.33,.6,.3,.3]),plot(r1,sigmat1,r1,sigmar1,r1,sigmaeq1,r1,sigmayp,'k')
;
title('   stresses at p max'); xlabel('barrel thickness');ylabel('stress');
gtext('tensile stress');gtext('radial stress');gtext('equivalent stress');gtext('yield stress')
subplot('position',[.1,.1,.3,.3]),plot(r2,sigmat2,r2,sigmar2,r2,sigmaeq2,r2,sigmayp,'k');
title('   stresses at mid section'); xlabel('barrel thickness');ylabel('stress');
gtext('tensile stress');gtext('radial stress');gtext('equivalent stress');gtext('yield stress')
subplot('position',[.6,.1,.3,.3]),plot(r3,sigmat3,r3,sigmar3,r3,sigmaeq3,r3,sigmayp,'k');
title('   stresses at muzzle end'); xlabel('barrel thickness');ylabel('stress');
gtext('tensile stress');gtext('radial stress');gtext('equivalent stress');gtext('yield
stress')

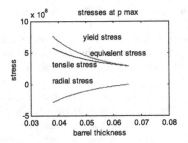

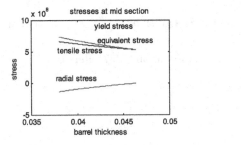
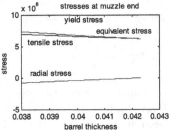

Fig 1.5.7: Stresses in the gun barrel at chamber, mid section and muzzle

1.6 Autofrettaged Gun Barrel Design

Basis of Autofrettage

In any gun barrel, the inner diameters of both the bore and the chamber are fixed by the projectile dimensions and the interior ballistics requirement. Hence the variables in the design of gun barrels are the wall thickness and the yield strength of the material proposed to be used.

In monobloc barrel construction, the wall thickness is dictated by the ratio of internal pressure to the yield strength of the material. With reference to Equation [1.4.6] it is seen that for $\frac{p_i}{\sigma_a}$ ratios of greater than 0.57735, the wall ratio is infinite. In the design of a barrel, the internal pressure is fixed by ballistics considerations; hence the designer has to resort to pre-stressing of the gun barrel in order to accommodate stresses in excess of the allowable stress of a given material.

In the autofrettage method of pre-stressing of gun barrels, the inner portion of the barrel wall is stressed beyond the elastic limit. However the yield is not exceeded at the outer surface. On removal of the autofrettage pressure, the outer surface, which is elastic, compresses the inner portion resulting in considerable residual compressive stress.

The general expression for autofrettage pressure is given by:

$$p_{af} = -2\tau_{max}\ln\frac{a}{b} + \frac{\tau_{max}\left(c^2 - b^2\right)}{c^2} \quad \text{.................................. Equation [1.5.7], Reference 1}$$

τ_{max} : maximum shear stress at yield
a: barrel inner radius
b: interface radius
c: barrel outer radius

If the plastic–elastic interface is at the outer radius:

$$p_{af} = 2\tau_{max} \ln\frac{c}{a}$$

From the Maximum Shear Stress Theory:

$$\sigma_{yp} = 2\tau_{max}$$

Hence:

$$p_{af} = \sigma_{yp} \ln\frac{c}{a} \quad\text{..} \quad [1.6.1]$$

σ_{yp} : yield strength of barrel material

If under the autofrettage pressure, yield in the barrel wall just reaches the outer surface, the residual hoop stress at the outer surface of the barrel is given by the algebraic sum of the yield stress and the hoop stress due to autofrettage pressure:

$$\sigma_{t(res)e} = \sigma_{yp} - \frac{a^2 p_{af}}{c^2 - a^2}\left(1 + \frac{c^2}{c^2}\right)$$

Here subscript e denotes the elastic condition

This in terms of the wall ratio T is:

$$\sigma_{t(res)e} = \sigma_{yp} - \frac{2p_{af}}{T^2 - 1} \quad\text{..} \quad [1.6.2]$$

Here:

$$T = \frac{c}{a} \quad\text{..} \quad [1.6.3]$$

T : wall ratio of the autofrettaged barrel

The working hoop stress at the outer surface is the sum of the residual hoop stress and the hoop stress due to propellant gas pressure:

$$\sigma_{t(work)e} = \sigma_{t(res)e} + \frac{a^2 p_i}{c^2 - a^2}\left(1 + \frac{c^2}{c^2}\right)$$

p_i: PIMP

Or:

$$\sigma_{t(work)e} = \sigma_{t(res)e} + \frac{2p_i}{T^2 - 1} \quad\dots\dots\dots\dots\dots\dots\dots\dots\dots\dots\dots\dots\dots\dots\dots\dots\dots\dots\dots\dots\dots\dots \text{[1.6.4]}$$

Therefore, combining Equations [1.6.3] and [1.6.4], the working hoop stress at the outer surface is:

$$\sigma_{t(work)e} = \sigma_{yp} - \frac{2p_{af}}{T^2 - 1} + \frac{2p_i}{T^2 - 1}$$

Finally:

$$\sigma_{t(work)e} = \sigma_{yp} - \left(p_{af} - p_i\right)\frac{2}{T^2 - 1} \quad\dots\dots\dots\dots\dots\dots\dots\dots\dots\dots\dots\dots\dots\dots\dots\dots\dots\dots\dots \text{[1.6.5]}$$

Evident from Equation [1.6.4] is the fact that the working hoop stress in the barrel material will exceed the yield strength of the material only if the maximum propellant gas pressure exceeds the autofrettage pressure. In other words, the barrel is safe for all pressures not in excess of the autofrettage pressure.

Methods of Autofrettage

Autofrettage is performed by two basic methods, hydraulic and mandrel also known as swage autofrettage.

Hydraulic Autofrettage

In hydraulic autofrettage yield in the material of the barrel wall is achieved by applying a high uniform hydraulic pressure inside the complete bore. Hydraulic autofrettage may be done in one of two ways. The first method is one in which the rough forging is unsupported from the outside. In the second case, the rough forging is supported from the outside.

Unsupported Hydraulic Autofrettage

As mentioned, this method involves the introduction of a uniform hydraulic pressure into the bore of an unsupported rough barrel forging. The term rough forging here implies that the inner radius of the forging is less than the final bore radius to cater for permanent enlargement during the autofrettage process and final machining after the autofrettage, whilst the outer radius is more than the final finished outer radius, allowance being made for machining. A parallel forging is used when the chamber diameter is equal to that of the bore. In case there is a significant difference between the two dimensions, the forging is prepared so that it is thick around the chamber portion and part of the inside of the forging is machined out to conform approximately to the chamber dimensions. The final forging has a uniform wall ratio that allows for the application of a uniform hydraulic autofrettage pressure. As the wall ratio is constant, yield will reach all points on the outer surface of the forging at the same time.

Container Method of Autofrettage

The second method involves the use of a container into which the rough forging is inserted and which supports the forging during the duration of introduction of hydraulic pressure into the bore. The container is a cylinder of adequate thickness to resist any permanent deformation during the autofrettage process. The interior of the container conforms closely to the finished profile of the barrel, while on the outside it is cylindrical. The clearance between the inner surface of the container and the outer surface of the forging is carefully calculated so that expansion of the external surface of the forging is kept strictly within its elastic limit. The container thus serves the purpose of allowing permanent enlargement of the bore, by introduction of uniform high pressure, yet retaining the outer surface within its elastic limit.

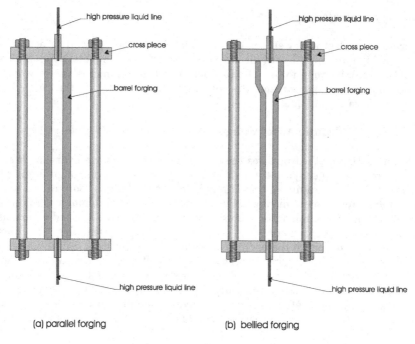

(a) parallel forging   (b) bellied forging

Fig 1.6.1: Unsupported autofrettage

The basic container autofrettage apparatus is shown diagrammatically in Fig 1.6.1. The barrel forging is placed inside the container suitably centered. The high pressure is developed in two or more stages. The first stage consists of a hydraulic pump, gauge and fluid lines. The output from the first stage is fed to the second stage which intensifies the pressure further. Accurate gauges are used to measure the pressure inside the bore. On attainment of the autofrettage pressure, calculated based on the maximum wall ratio, the pressure is released. A small clearance between the forging and the container, lesser than that which existed before the start of autofrettage, now develops between the two surfaces in question because of the relief of both container and barrel outer surface.

The forging is now inspected for post autofrettage defects before being subject to low heat treatment. This process is followed by cutting of the rifling grooves and final machining of the outer surface.

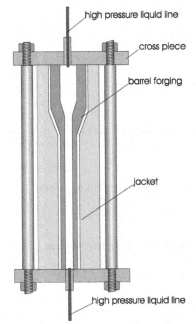

Fig 1.6.1: Container autofrettage apparatus

Empirical Formulae for Determination of Wall Ratio and ESP of Autofrettaged Gun Barrels

Equation [1.6.1], suitably adjusted, has been found by gun designers to be practically useful for the computation of wall ratio of autofrettaged gun barrels. The modified empirical equation is:

$$p_i = 0.9\sigma_a \, ln\frac{c}{a} \qquad\qquad\qquad [1.6.6]$$

σ_a : allowable stress

Equation [1.6.1] has been modified replacing the autofrettage pressure by the design pressure at the section. The yield strength of the material is reduced to an allowable stress and an additional safety factor of 0.9 is applied.

Equation [1.6.6] is now used for computation of the wall ratio as follows:

$$ln\frac{c}{a} = 1.111\frac{p_i}{\sigma_a}$$

Or:

$$T = e^{1.111\frac{p_i}{\sigma_a}} \dots\dots\dots [1.6.7]$$

Allowance for Eccentricity

Allowance for eccentricity is made as for monobloc barrels. Equation [1.3.5] refers.

Empirical Relation for ESP

Empirically, the ESP has been established according to the following expression:

$$ESP = 1.08\sigma_{yp}\ln T \dots\dots\dots [1.6.8]$$

Safety Factor

The safety factor is calculated as for monobloc barrels vide Equation [1.4.8].

Clearance between Barrel Outer Surface and Container Inner Surface before Autofrettage

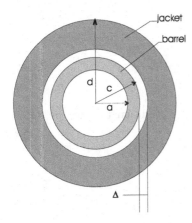

Fig 1.6.3: Initial clearance between barrel and container before autofrettage

The maximum clearance between the barrel and container should not exceed the maximum elastic displacement of the barrel outer surface under the autofrettage pressure. For every point in the plastic region of the barrel wall under full autofrettage pressure the following relation, according to the Maximum Shear Stress Theory, holds good:

$$\sigma_t - \sigma_r = \sigma_{yp} = 2\tau_{max} \quad \dots\dots\dots\dots\dots\dots\dots\dots\dots\dots\dots\dots\dots\dots\dots\dots\dots\dots\dots\dots\dots\dots\dots\dots\dots\dots [1.6.9]$$

σ_t : hoop stress

σ_r : radial stress

σ_{yp} : yield stress

τ_{max} : maximum shear stress at yield

From Equation [1.6.9]:

$$\sigma_r - \sigma_t = -\sigma_{yp}$$

Substituting in the Equation of Equilibrium $r \dfrac{d\sigma_r}{dr} + \sigma_r - \sigma_t = 0$:

$$\frac{d\sigma_r}{dr} = \sigma_{yp} \frac{1}{r}$$

Or:

$$\sigma_r = \sigma_{yp} \ln r + C$$

The constant of integration C is determined from the condition that at the outer radius of the barrel c, the radial stress $\sigma_r = -p_{if}$:

p_{if} : pressure at the barrel container interface

$$C = -\sigma_{yp} \ln c - p_{if}$$

It follows that:

$$\sigma_r = \sigma_{yp} \ln \frac{r}{c} - p_{if} \dotfill [1.6.10]$$

At the inner surface of the barrel, $\sigma_r = -p_{af}$ and $r = a$, hence:

$$- p_{af} = \sigma_{yp} \ln \frac{a}{c} - p_{if}$$

Or:

$$p_{if} = \sigma_{yp} \ln \frac{a}{c} + p_{af} \dotfill [1.6.11]$$

Displacement of the Outer Surface of the Barrel Wall

The circumferential strain of the barrel is given by:

$$\varepsilon_t = \frac{1}{E}(\sigma_t - \mu\sigma_r) \dots\dots\dots\dots\dots\dots\dots\dots\dots\dots\dots\dots\dots\dots\dots\dots\dots\dots\dots\dots \text{[1.6.12]}$$

μ : Poisson's Ratio

From which the change δ_{rb} in radius r of the barrel is:

$$\delta_{rb} = \frac{2\pi r\varepsilon_t}{2\pi} = r\varepsilon_t$$

Substituting for the circumferential strain from Equation [1.6.12]:

$$\delta_{rb} = \frac{r}{E}(\sigma_t - \mu\sigma_r) \dots\dots\dots\dots\dots\dots\dots\dots\dots\dots\dots\dots\dots\dots\dots\dots\dots\dots\dots\dots \text{[1.6.13]}$$

From Equations [1.6.9] and [1.6.10], the hoop stress in the barrel wall after autofrettage pressure is fully applied is:

$$\sigma_t = \sigma_{yp}\left(1 + ln\frac{r}{c}\right) - p_{if} \; ; \text{ at the outer radius } c \text{ of the barrel:}$$

$$\sigma_t = \sigma_{yp} - p_{if}$$

Applying this result in Equation [1.6.13], the inward deflection of the barrel outer surface due to the interface pressure is:

$$\delta_{rb} = \frac{c}{E}\left(\sigma_{yp} - p_{if} + \mu p_{if}\right)$$

Or:

$$\delta_{rb} = \frac{c}{E}\left[\sigma_{yp} - p_{if}(1 - \mu)\right] \dots\dots\dots\dots\dots\dots\dots\dots\dots\dots\dots\dots\dots\dots\dots\dots \text{[1.6.14]}$$

Displacement of the Inner Surface of the Container

The radial stress at the inner surface of the container is given by:

$$\sigma_{rc} = \frac{c^2 p_{if}}{d^2 - c^2}\left(1 - \frac{d^2}{c^2}\right)$$ .... Equation [1.2.12], Reference 1

c: inner radius of the container
d: outer radius of the container

It follows that:

$$\sigma_{rc} = -p_{if}$$

The hoop stress of the container at the interface is:

$$\sigma_{tc} = \frac{c^2 p_{if}}{d^2 - c^2}\left(1 + \frac{d^2}{c^2}\right)$$ .... Equation [1.2.13], Reference 1

Applying Equation [1.6.13], the deflection outward of the inner surface of the container due to the interface pressure is:

$$\delta_{rc} = \frac{c}{E_c}\left(\frac{c^2 p_{if}}{d^2 - c^2}\left\{\left(1 + \frac{d^2}{c^2}\right) - \mu_c\left(1 - \frac{d^2}{c^2}\right)\right\}\right) = \frac{c}{E_c}\left(\frac{p_{if}}{d^2 - c^2}\left\{\left(c^2 + d^2\right) + \mu_c\left(d^2 - c^2\right)\right\}\right)$$

Finally:

$$\delta_{rc} = \frac{c}{E_c}p_{if}\left(\frac{d^2 + c^2}{d^2 - c^2} + \mu_c\right)$$ .... [1.6.15]

μ_c: Poissons Ratio of the container material
E_c: Young's Modulus of the container material

The interference at the interface is:

$$\Delta = \delta_{rb} - \delta_{rc}$$ .... [1.6.16]

The inner radius of the container is hence:

$$r_{ci} = c + \Delta \dotfill [1.6.17]$$

Swage Autofrettage

Autofrettage may also done by forcing a mandrel through the bore of the rough forging. When a solid mandrel of radius greater than the inner radius of the barrel forging is forced through the bore, the effect produced in the material of the barrel is similar to that induced by hydraulic autofrettage.

Determination of Wall Ratio

The barrel wall ratio is determined, as in the case of hydraulic autofrettage, from Equation [1.6.7]:

$$T = e^{1.111\frac{p_i}{\sigma_{yp}}}$$

The barrel wall ratio having been determined it is now possible to compute the autofrettage pressure which must be generated by the forcing of the oversize mandrel through the bore to cause yield just up to the outer surface by making use of Equation [1.6.1]:

$$p_{af} = \sigma_{yp} \ln\frac{c}{a}$$

c: barrel outer radius
a: barrel inner radius

Estimation of Mandrel Radius

The radius of the mandrel determines the autofrettage pressure and hence the distance radially outwards to which yield will take place in the barrel material, as also the final and desired inner radius of the bore. The mandrel radius is calculated to take into account both the displacement of the material of the barrel and the mandrel itself.

Displacement of Inner Surface of the Barrel

Consider a gun barrel of incompressible material and of inner radius a and outer radius c subject to an internal pressure p_i. The barrel is restrained along its length so that the axial strain $\varepsilon_a = 0$. If u is the radial displacement of a cylindrical surface of radius r, within the barrel wall, then the displacement of a cylindrical surface of radius $r + dr$ is $u + \dfrac{du}{dr} dr$. Hence an element of radial thickness dr undergoes a total elongation in the radial direction of $\dfrac{du}{dr} dr$ and the radial strain is given by:

$$\varepsilon_r = \frac{du}{dr} \quad \text{..} \quad [1.6.18]$$

The tangential strain is given by the radial strain of the corresponding radius:

$$\varepsilon_t = \frac{u}{r} \quad \text{..} \quad [1.6.19]$$

The material of the barrel is deemed incompressible; hence the volumetric strain is zero as expressed by the following condition:

$$\varepsilon_r + \varepsilon_t + \varepsilon_a = 0$$

As the axial strain is zero:

$$\varepsilon_r + \varepsilon_t = 0 \quad \text{..} \quad [1.6.20]$$

Substituting for $\varepsilon_r, \varepsilon_t$ from Equations [1.6.18] and [1.6.19] in Equation [1.6.20]:

$$\frac{du}{dr} + \frac{u}{r} = 0$$

Integrating:

$$\ln u = -\ln r + \ln C$$

44

From which:

$$u = \frac{C}{r}$$

The constant C is determined from the condition at the inner surface of the barrel, where $r = a$ and the displacement is u_a. Hence:

$$C = u_a a$$

The displacement of a cylindrical surface of radius r is:

$$u = \frac{u_a a}{r} \quad \text{.. [1.6.21]}$$

It follows that the radial and tangential strains can be expressed as:

$$\varepsilon_r = \frac{du}{dr} = -\frac{u_a a}{r^2} \quad \text{.. [1.6.22]}$$

$$\varepsilon_t = \frac{u}{r} = \frac{u_a a}{r^2} \quad \text{.. [1.6.23]}$$

Condition of Pure Shear

If the stresses with reference to the three principal axis taken in the radial, tangential and axial directions of the barrel be σ_r, σ_t and σ_a respectively, it is then possible to superimpose a third stress equal to $\frac{\sigma_r + \sigma_t + \sigma_a}{3}$ on each of the stresses so that the stresses in the radial, tangential and axial directions respectively may be separated into the components contributing purely to deformation and components contributing purely to change in volume. The stress components contributing to deformation are:

$$\sigma_r - \frac{\sigma_r + \sigma_t + \sigma_a}{3} = \frac{2\sigma_r - \sigma_t - \sigma_a}{3}$$

$$\sigma_t - \frac{\sigma_r + \sigma_t + \sigma_a}{3} = \frac{2\sigma_t - \sigma_r - \sigma_a}{3}$$

$$\sigma_a - \frac{\sigma_r + \sigma_t + \sigma_a}{3} = \frac{2\sigma_a - \sigma_r - \sigma_t}{3} .$$

The stress components contributing to purely volumetric change are:

$$\frac{\sigma_r + \sigma_t + \sigma_a}{3}, \quad \frac{\sigma_r + \sigma_t + \sigma_a}{3} \text{ and } \frac{\sigma_r + \sigma_t + \sigma_a}{3} .$$

Considering the case of pure shear:

The radial strain is given by:

$$\varepsilon_r = \frac{2\sigma_r - \sigma_t - \sigma_a}{3E} - \mu \frac{2\sigma_t - \sigma_r - \sigma_a}{3E} - \mu \frac{2\sigma_a - \sigma_r - \sigma_t}{3E}$$

Or:

$$\varepsilon_r = \frac{1 + \mu}{3E} \left(2\sigma_r - \sigma_t - \sigma_a \right)$$

∵ The Modulus of Rigidity G is related to Young's Modulus E

by the relation $G = \dfrac{E}{2(1 + \mu)}$:

$$\left(2\sigma_r - \sigma_t - \sigma_a \right) = 6G\varepsilon_r \quad \text{...} \quad [1.6.24]$$

Similarly in the tangential direction:

$$\left(2\sigma_t - \sigma_r - \sigma_a \right) = 6G\varepsilon_t$$

ε_t : tangential strain

Applying Equation [1.6.20]:

$$\left(2\sigma_t - \sigma_r - \sigma_a\right) = -6G\varepsilon_r \quad\text{.. [1.6.25]}$$

Finally in the axial direction:

$$\left(2\sigma_a - \sigma_r - \sigma_t\right) = 0 \quad\text{.. [1.6.26]}$$

Combining Equations [1.6.24] and [1.6.25]:

$$\sigma_r - \sigma_t = 4G\varepsilon_r$$

Making use of Equation [1.6.22]:

$$\sigma_r - \sigma_t = -4G\frac{u_a a}{r^2}$$

Within elastic limits the radial and tangential stresses in a gun barrel can be expressed by the following equations:

$$\sigma_r = \frac{p_i a^2}{c^2 - a^2}\left(1 - \frac{c^2}{r^2}\right) \quad\text{..[1.2.12] Reference 1}$$

And:

$$\sigma_t = \frac{p_i a^2}{c^2 - a^2}\left(1 + \frac{c^2}{r^2}\right) \quad\text{..[1.2.13] Reference 1}$$

It may be noted that here the outer radius of the barrel denoted by b in Reference 1 is here denoted by c.

The axial stress is determined from the force due to the axial pressure p_i acting on a cross section of the barrel divided by the area of the cross section:

$$\sigma_a = \frac{p_i a^2}{c^2 - a^2} \quad\text{.. [1.6.27]}$$

47

Substituting for σ_r, σ_t from Equations [1.2.12] & [1.2.13] Reference 1:

$$u_a = \frac{p_i}{2G} \frac{a}{1 - \dfrac{a^2}{c^2}} \quad\quad\dotfill\quad [1.6.28]$$

The Von Mises–Hinds–Hencky criterion for yielding is given by:

$$\left(\sigma_r - \sigma_t\right)^2 + \left(\sigma_t - \sigma_a\right)^2 + \left(\sigma_a - \sigma_r\right)^2 = 2\sigma_{yp}^2$$

Substituting for $\sigma_r, \sigma_t, \sigma_a$ from Equations [1.2.12] & [1.2.13] Reference 1 and Equation [1.6.27] respectively, the internal pressure which will cause yield to take place at a radius r if the material is of yield strength σ_{yp} is given by:

$$p_r = \frac{\sigma_{yp}}{\sqrt{3}} \frac{1 - \dfrac{a^2}{c^2}}{\dfrac{a^2}{r^2}} \quad\quad\dotfill\quad [1.6.29]$$

At the inner surface $r = a$, hence:

$$p_a = \frac{\sigma_{yp}}{\sqrt{3}}\left(1 - \frac{a^2}{c^2}\right) \quad\quad\dotfill\quad [1.6.30]$$

As the internal pressure p_r is increased from p_a to p_{af}, the former being the pressure necessary to cause yield at the inner surface of the barrel and the latter the autofrettage pressure necessary to cause yield up to the outer radius c, the radius of the plastic-elastic interface, say b, increases from a to c. At the plastic elastic interface $r = b$. Also the inner radius of the elastic region is b, so it follows that the pressure which will cause yield up to the interface is:

$$p_b = \frac{\sigma_{yp}}{\sqrt{3}}\left(1 - \frac{b^2}{c^2}\right) \quad\quad\dotfill\quad [1.6.31]$$

Applying Equation [1.6.28], and substituting $p_r = p_b$ and $a = b$, the displacement of all points on the plastic elastic interface is given by:

$$u_b = \frac{p_b}{2G} \frac{b}{1 - \dfrac{b^2}{c^2}}$$

Applying Equation [1.6.31], the displacement at the interface is:

$$u_b = \frac{\sigma_{yp}}{\sqrt{3}} \left(1 - \frac{b^2}{c^2}\right) \frac{1}{2G} \frac{b}{1 - \dfrac{b^2}{c^2}}$$

Finally:

$$u_b = \frac{\sigma_{yp} b}{2\sqrt{3}G} \quad \text{..} \quad [1.6.32]$$

The displacement at the interface is related to the displacement at the inner surface by Equation [1.6.21]:

$$u_b = \frac{u_a a}{b}$$

Hence the displacement at the inner surface is given by:

$$u_a = \frac{\sigma_{yp} b}{2\sqrt{3}G} \frac{b}{a}$$

Or:

$$u_a = \frac{\sigma_{yp} b^2}{2\sqrt{3}Ga} \quad \text{..} \quad [1.6.33]$$

At the beginning of autofrettage $b = a$, hence:

$$u_a = \frac{\sigma_{yp} a}{2\sqrt{3}G}$$

At the end of autofrettage $b = c$, hence:

$$u_a = \frac{\sigma_{yp} c^2}{2\sqrt{3} G a} \qquad [1.6.34]$$

Radial Displacement of the Mandrel

From established formulae, when a solid cylinder, such as the mandrel, is subject to external pressure alone, in this case the autofrettage pressure, the radial displacement of the mandrel is given by:

$$u_m = P_{af} \frac{a}{E_m} (1 - \mu_m) \qquad [1.6.35]$$

a: inner radius of barrel and nominated radius of mandrel
E_m: Young's Modulus of mandrel material
μ_m: Poissons Ratio of mandrel material

The necessary radial interference between the barrel and mandrel is:

$$i = u_a + u_m \qquad [1.6.36]$$

The final mandrel radius is hence:

$$r_m = a + i \qquad [1.6.37]$$

Finishing of Autofrettaged Gun Barrels

Subsequent to autofrettage, the barrel is machined to its final exact dimensions. In order that the beneficial effect of prestressing be preserved to the largest extent possible, the dimensions of the rough forging must conform as closely as possible to the dimensions which after the pre-stressing process will be the final required dimensions.

1.7 Design of an Autofrettaged Gun Barrel

Design Exercise 1.7.1

The preliminary design, including allowance for eccentricity, of an autofrettaged gun barrel, similar in ballistics to that described in Article 1.5, is illustrated here. The minimum wall thicknesses in the case of monobloc and autofrettage gun barrels under similar ballistics conditions are compared. Also the possible reduction in yield strength of material used, if the wall thickness of the autofrettaged barrel is identical to that of the monobloc barrel under the same condition of maximum gas pressure is demonstrated.

The theoretical internal dimensions of chamber and bore are as given in Fig 1.5.2. Interior ballistics data is contained in the worksheet of Fig 1.5.1.

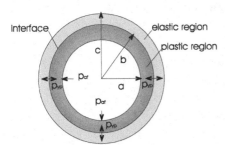

Fig 1.7.1: Cross section through a gun barrel under autofrettage pressure

Design Procedure

Computation of Wall Thickness

Wall thickness is computed from the gun design curve with the help of Equations [1.6.3] and [1.6.7]

ESP

ESP is computed with the help of Equation [1.6.8] inserting the value of the final wall ratio after making allowance for eccentricity during manufacture is made by the application of Equation [1.3.5], as in the case of monobloc gun barrels.

Safety Factor

The Safety Factor is computed with the help of Equation [1.3.4], also as in the case of monobloc gun barrels.

Matlab Programme 1.7.1

```
% programme to compute wall thickness, ESP, SF against barrel length for af barrel
% file name: af.txt
sigma_yp=841.*10^6 % yield strength
sigmaa=772.05.*10^6 % allowable stress
range1='h7..h28'
p_i=wk1read('a:pimp',6,7,range1) % design pressure
range2='e7..e28'
l=wk1read('a:pimp',6,4,range2) % barrel elngth
T=exp(1.111.*p_i./sigmaa) % wall thickness ratio
range3='g7..g28'
r_i=wk1read('a:pimp',6,6,range3) % inner radius
r_o=T.*r_i % outer radius
T_adj=1.0526.*T-.0526 % adjusted wall ratio
ESP=sigma_yp.*(T_adj.^2-1)./sqrt(3.*T_adj.^4+1)
SF=ESP./p_i
subplot('position',[.1,.4,.3,.4]),plot(l,ESP);title('        (a) barrel length vs
ESP');xlabel('barrel length');ylabel('ESP')
subplot('position',[.6,.4,.3,.4]), plot(l,SF);title('    (b) barrel length vs safety
factor');xlabel('barrel length');ylabel('safety factor')
subplot('position',[.1,.1,.8,.15]),plot(l,r_i,l,r_o);title('(c) barrel length vs wall
thickness');xlabel('barrel length');ylabel('wall thickness')
```

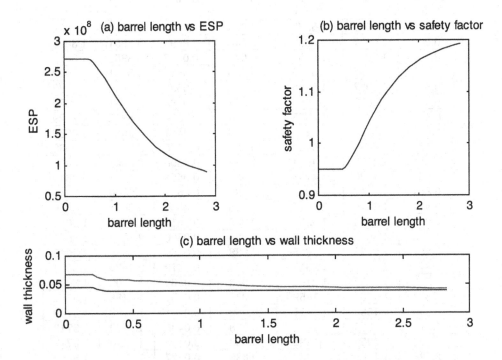

Fig 1.7.2: Graphical results of autofrettaged gun barrel design.

Barrel Length m	Wall Ratio m	Adjusted Wall Ratio	Inner Radius m	Outer Radius	ESP $*10^8$ Pa	SF
0.000	1.5087	1.5355	0.045	0.0691	3.8951	1.3629
0.200	1.5087	1.5355	0.045	0.0691	3.8951	1.3629
0.238	1.5087	1.5355	0.042	0.0637	3.8951	1.3629
0.295	1.5087	1.5355	0.038	0.0583	3.8951	1.3629
0.351	1.5087	1.5355	0.038	0.0583	3.8951	1.3629
0.408	1.5087	1.5355	0.038	0.0583	3.8951	1.3629
0.464	1.5087	1.5355	0.038	0.0583	3.8951	1.3629
0.521	1.5025	1.5289	0.038	0.0581	3.8563	1.3630
0.578	1.4890	1.5148	0.038	0.0576	3.7716	1.3633
0.634	1.4707	1.4954	0.038	0.0568	3.6551	1.3636
0.691	1.4520	1.4758	0.038	0.0561	3.5351	1.3640
0.804	1.4117	1.4334	0.038	0.0545	3.2700	1.3647
0.917	1.3720	1.3915	0.038	0.0529	3.0011	1.3656
1.030	1.3354	1.3530	0.038	0.0514	2.7461	1.3663
1.143	1.3002	1.3160	0.038	0.0500	2.4944	1.3671
1.257	1.2722	1.2865	0.038	0.0489	2.2885	1.3678
1.370	1.2481	1.2611	0.038	0.0479	2.1074	1.3684
1.596	1.2057	1.2165	0.038	0.0462	1.7802	1.3695
1.823	1.1766	1.1859	0.038	0.0451	1.5488	1.3702
2.049	1.1530	1.1611	0.038	0.0441	1.3563	1.3709
2.275	1.1375	1.1448	0.038	0.0435	1.2279	1.3713
2.502	1.1256	1.1322	0.038	0.0430	1.1280	1.3717
2.728	1.1168	1.1229	0.038	0.0427	1.0531	1.3719
2.825	1.1136	1.1196	0.038	0.0425	1.0261	1.3720

Fig 1.7.3: Tabulated results of autofrettaged gun barrel design.

Autofrettage Pressure

Having established the adjusted wall ratio, it is now possible to calculate the autofrettage pressure necessary to cause yield up to the outer surface at the section of maximum thickness.

From Equation [1.6.1]:

$$p_{af} = 841.10^6 \, ln \, 1.5355 = 360.066 \, \text{M Pa}$$

Design Exercise 1.7.2

Stress Analysis of Autofrettaged Gun Barrel

Stress analysis of the autofrettaged gun barrel designed in 1.7 is carried out at the position of maximum gas pressure. From Fig 1.7.2, the inner and outer radius at the position of maximum gas pressure is 0.038 and 0.0583 respectively. The PIMP from Fig 1.5.3 is 286 M Pa. The maximum shear stress at yield is 420.5 M Pa. Keeping in mind that the plasticity is induced just up to the outer surface, the interface radius equals the outer radius. The stresses are computed as follows:

Stresses due to Deformation

Radial stress is given by the expression:

$$\sigma_{r1} = 2\tau_{max} \, ln \, \frac{r}{b}$$

Hoop stress is given by:

$$\sigma_{t1} = 2\tau_{max} + \sigma_{r1}$$

Stresses due to Autofrettage Pressure

Autofrettage pressure is given by:

$$p_{af} = -2\tau_{max} \, ln \, \frac{a}{b}$$

Radial stress is given by:

$$\sigma_{r2} = \frac{a^2 p_{af}}{(c^2 - a^2)} \left(1 - \frac{c^2}{r^2} \right)$$

55

Hoop stress is given by:

$$\sigma_{t2} = \frac{a^2 p_{af}}{(c^2 - a^2)} \left(1 + \frac{c^2}{r^2} \right)$$

Residual Stresses

Radial residual stress is:

$$\sigma_{r3} = \sigma_{r1} - \sigma_{r2}$$

Residual hoop stress is:

$$\sigma_{t3} = \sigma_{t1} - \sigma_{t2}$$

Stresses due to Propellant gas Pressure

Radial stress is given by:

$$\sigma_{r4} = \frac{a^2 p_{max}}{(c^2 - a^2)} \left(1 - \frac{c^2}{r^2} \right)$$

Hoop stress is obtained from:

$$\sigma_{t4} = \frac{a^2 p_{af}}{(c^2 - a^2)} \left(1 + \frac{c^2}{r^2} \right)$$

Working Stresses

Radial working stress is:

$$\sigma_{r5} = \sigma_{r3} + \sigma_{r5}$$

Working hoop stress is:

$$\sigma_{t5} = \sigma_{t3} + \sigma_{t5}$$

Equivalent Stress

The equivalent stress is obtained from the Mises-Hencky expression for a two-dimensional stress condition:

$$\sigma_{eq} = \sqrt{\sigma_{r5}^2 - \sigma_{r5}\sigma_{t5} + \sigma_{t5}^2}$$

The stresses are analyzed based on the above expressions with the help of a Matlab programme given below:

Matlab Programme 1.7.2

```
% programme to compute stresses in af barrel
% file name: af stress

% data
a=.038 % inner radius m
b=.0583 % outer radius m
c=b

r=linspace(a,b,100) % radius array
r1=linspace(a,b,1000) % radius array

towmax=420.5*10^6 % max shear stress Pa
syp=2.*towmax % yield stress % yield stress Pa
pmax=286*10^6 % max propellant gas pressure

% stress dt deformation

% plastic region
sr1=2.*towmax.*log(r./b)-towmax.*(c.^2-b.^2)./c.^2
st1=2.*towmax +sr1

% pyp=0 at b=c

% stresses dt afpressure
```

```
paf=-2.*towmax.*log(a./b)
sr2=a.^2.*paf./(c.^2-a.^2).*(1-c.^2./r.^2)
st2=a.^2.*paf./(c.^2-a.^2).*(1+c.^2./r.^2)

% residual stresses
sr3=sr1-sr2
st3=st1-st2

% stress dt gas pressure
sr4=a.^2.*pmax./(c.^2-a.^2).*(1-(c.^2./r.^2))
st4=a.^2.*pmax./(c.^2-a.^2).*(1+(c.^2./r.^2))

% working stresses
sr5=sr3+sr4
st5=st3+st4

% equivalent stress
seq=sqrt(sr5.^2 + st5.^2-sr5.*st5)

plot(r1,syp,r,sr5,r,st5,r,seq)
title('Stresses in autofrettaged barrel')
xlabel('radius');ylabel('stress')
gtext('yield stress')
gtext('working radial  stress')
gtext('working hoop  stress')
gtext('equivalent stress')
```

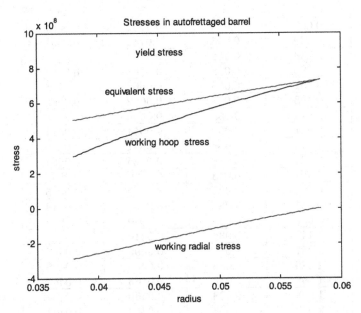

Fig 1.7.4: Yield, working and equivalent stresses in an autofrettaged gun barrel

Barrel length m	Inner radius m	Outer radius monobloc m	Outer radius autofrettage m
0.000	0.045	0.0781	0.0691
0.200	0.045	0.0781	0.0691
0.238	0.042	0.0721	0.0637
0.295	0.038	0.0661	0.0584
0.351	0.038	0.0662	0.0585
0.408	0.038	0.0662	0.0585
0.464	0.038	0.0660	0.0583
0.521	0.038	0.0654	0.0581
0.578	0.038	0.0641	0.0576
0.634	0.038	0.0625	0.0568
0.691	0.038	0.0610	0.0561
0.804	0.038	0.0580	0.0545
0.917	0.038	0.0553	0.0529
1.030	0.038	0.0530	0.0514
1.143	0.038	0.0510	0.0500
1.257	0.038	0.0495	0.0489
1.370	0.038	0.0483	0.0479
1.596	0.038	0.0463	0.0462
1.823	0.038	0.0450	0.0451
2.049	0.038	0.0440	0.0441
2.275	0.038	0.0433	0.0435
2.502	0.038	0.0428	0.0430
2.728	0.038	0.0424	0.0427
2.825	0.038	0.0423	0.0425

Fig 1.7.4: Difference between outer radii of monobloc and autofrettaged gun barrels of Design Exercises 1.5.1 & 1.7.1

Comparison between Monobloc and Autofrettage: Reduction in Yield Strength

As will be illustrated, a significant reduction in the yield strength of the material can be accepted in the case of an autofrettaged gun barrel if the wall thicknesses of a monobloc barrel designed for similar ballistics are accepted.

The maximum design pressure is obtained from Fig 1.5.3.

$p_{i(max)} = 2.86.10^8$ M Pa

The thickness ratio at the position of maximum gas pressure is obtained from Fig 1.5.6:

Wall thickness ratio of the monobloc barrel: $T = 1.7365$

Substituting in Equation [1.6.7]:

$$ln\,1.7365 = 1.111\frac{286.10^6}{\sigma_a}$$

From which:

$\sigma_a = 575$ M Pa

Allowable stress of steel earlier used was 772.05 M Pa. Hence the reduction in yield strength possible for barrels of similar dimensions by the monobloc and autofrettage processes is significantly 197.05 M Pa.

Design Exercise 1.7.3

Calculation of Barrel-Container Clearance

Calculation of the desired clearance between the barrel and container for the hydraulic autofrettage of the gun barrel in Design Exercise 1.7.1 and the inner radius of the container, given that:

Wall ratio of the container is 3.0

E for the steel of both container and barrel is 210.00 G Pa

Poisson's Ratio for container and barrel is 0.3

Yield strength of the container material is 0.4 G Pa.

Autofrettage Pressure

The autofrettage pressure at full autofrettage is calculated for the maximum thickness ratio using Equation [1.6.1] and the actual yield strength of the material.

$$p_{af} = 841.0.10^6 \, ln \, 1.5355 = 360.67.10^6 \, Pa$$

Pressure at the Barrel-Container Interface

Pressure at the barrel container interface as given by Equation [1.6.11] is:

$$p_{if} = \sigma_{yp} \, ln \, \frac{a}{c} + p_{af} = 841.0.10^6 \, ln \, \frac{0.038}{0.0583} + 360.67.10^6 = 0.7086 \; M \, Pa$$

Displacement of the Outer Surface of the Barrel

Displacement of the outer surface of the barrel inwards due to the interface pressure is given by Equation [1.6.14]:

$$\delta_r = \frac{c}{E} \left[\sigma_{yp} - p_{if}(1 - \mu) \right] = \frac{0.0583}{210.10^9} \left[841.10^6 - 0.7086.10^6 (1 + 0.3) \right] = 0.233 \, mm$$

Displacement of Container Due to Interface Pressure

Outward radial displacement of the container inner surface due to the interface pressure is given by Equation [1.6.15]

$$\delta_{rc} = \frac{c}{E_c} \, p_{if} \left(\frac{d^2 + c^2}{d^2 - c^2} + \mu_c \right) = \frac{0.0583}{210.10^9} \, 0.7086.10^6 \left(\frac{3^2 + 1}{3^2 - 1} - 0.3 \right) = 3.049.10^{-4} \, mm$$

Vide Equation [1.6.16], the interference between barrel and container is:

$$\Delta = \delta_{rb} - \delta_{rc} = 0.233 - 3.049.10^{-4} = 0.2327 \text{ mm}$$

Hence the desired inner radius of the container as per Equation [1.6.17] is:

$$r_{ic} = 0.05853 \text{ m}$$

Design Exercise 1.7.4

Design of a Gun Barrel by Swage Autofrettage

A 100 mm gun barrel is to be designed by the method of swage autofrettage. The steel selected for the forging is AISI 4150 carbon steel in the quenched and tempered condition, of tensile strength 1207 MPa and yield strength 1103 MPa. The Modulus of Rigidity is 81.36 G Pa, Young's modulus is 210 G Pa and Poisson's Ratio 0.293. The computed maximum pressure is 294.2 MPa. Compute the wall ratio, the autofrettage pressure and the mandrel radius for the position of maximum gas pressure.

The design pressure as given by Equation [1.3.2]:

$$p_i = 1.15CMP + 19.302 = 1.15.294.2 + 19.302 = 357.632 \text{ M Pa}$$

The allowable stress of the barrel steel is given by Equation [1.3.3]:

$$\sigma_a = 1103.0 - 68.95 = 1034.05 \text{ M Pa}$$

The barrel wall ratio is determined from Equation [1.6.7]:

$$T = e^{1.111\frac{p_i}{\sigma_a}} = e^{1.111\frac{357.632}{1034.05}} = 1.4685$$

The adjusted wall ratio vide Equation [1.3.5] is:

$$T_{adj} = 1.0526.1.4685 - 0.0526 = 1.493$$

Outer radius of the barrel wall is 74.65 mm

The barrel wall ratio having been determined it is now possible to compute the autofrettage pressure necessary to cause yield just up to the outer surface by making use of Equation [1.6.1]:

$$p_{af} = \sigma_{yp} \, ln \frac{c}{a} = 1103 \, ln \, 1.493 = 442.07 \text{ M Pa}$$

Displacement of the inner surface of the barrel is given by Equation [1.6.34]:

$$u_a = \frac{\sigma_{yp} c^2}{2\sqrt{3}Ga} = \frac{1103.0.74.65^2}{2\sqrt{3}.81360.0.50.0} = 0.436 \text{ mm}$$

Radial displacement of the mandrel as given by Equation [1.6.35]:

$$u_m = p_{af} \frac{a}{E} (1-\mu) = 442.07 \frac{50}{210000} (1 - 0.293) = 0.0744 \text{ mm}$$

Radial interference from Equation [1.6.36] is:

$$i = 0.436 + .0744 = 0.5104 \text{ mm}$$

Mandrel radius as given by Equation [1.6.33]:

$$r_m = a + i = 50.0 + 0.5104 = 50.5104 \text{ mm}$$

1.8 Rifling Design

Rifling Design Considerations

The basis of effective rifling design may be summarized as follows. The correct spin rate must be imparted to the projectile as it leaves the muzzle. External forces acting on the projectile should be minimum at the instant of its exit from the muzzle. Stresses on both barrel and the driving band of the projectile, during the phase of interaction of the projectile with the rifling, must be again the least possible. The centrifugal force produced due to the rifling should be sufficient for efficient removal of safety devices and arming of the fuze system. Finally the system of rifling and driving band must be simple to manufacture and employ.

Definitions Associated with Rifling Design

Driving Edge of Groove. This is the edge of the groove, which is in firm contact with the driving band as the projectile moves up the barrel.

Windage: This is the clearance between the projectile and the bore.

Uniform Twist of Rifling: Rifling in which the angle of rifling is a constant.

Increasing Twist of Rifling: Rifling in which the angle of rifling increases towards the muzzle.

Combined Twist of Rifling: A combination of both uniform and increasing twist.

Groove Root: The point where the groove walls and base of the groove meet.

Design Parameters of Rifling

Rifling design involves the definition of the following two main parameters of rifling, namely the type of twist and the rifling profile.

Type of Twist

Stability of spun projectiles is determined by the rate of spin at the instant of projectile exit from the barrel. Therefore, the barrel is designed with the angle of rifling at the muzzle in accordance with this requirement. However, from the origin of rifling up to the point of the muzzle, considerable flexibility is available to the designer by way of choice of the rifling curve. As earlier seen, the rifling curve may be uniform, parabolic, cubic, semi cubic or the increase in the angle of rifling may be according to some exponent n.

In general all rifling curves conform to the expression:

$$y = Kx^n \dotfill [1.8.1]$$

n is the exponent defining the rifling curve

$n = 1$: uniform twist

$n = 1.5$: semi cubic twist

$n = 2$: parabolic twist

$n = 3$: cubic twist

n may also assume any value between 1 and usually not more than 3

x: projectile displacement in the axial direction of the bore

y: peripheral displacement of a point on the projectile circumference corresponding to axial displacement x.

K: a constant which is found from the angle of the rifling at exit

Differentiating Equation [1.8.1]:

$$\frac{dy}{dx} = Knx^{n-1}$$

The slope of the rifling curve:

$$\frac{dy}{dx} = Tan\,\alpha$$

α : angle of rifling at the point under consideration

Considering the point of exit at which the angle of rifling is α_E and the projectiles displacement in the axial direction is x_E:

$$\left(\frac{dy}{dx}\right)_E = Tan\,\alpha_E = Knx_E^{n-1}$$

From which:

$$K = \frac{Tan\,\alpha_E}{nx_E^{n-1}}$$

Equation [1.8.1] may be rewritten as:

$$y = \frac{Tan\,\alpha_E}{nx_E^{n-1}}\,x^n$$

It follows that the slope of the rifling curve is given by:

$$\frac{dy}{dx} = \frac{Tan\,\alpha_E}{x_E^{n-1}}\,x^{n-1} \quad\text{..} \quad [1.8.2]$$

And the rate of change of the rifling slope is:

$$\frac{d^2y}{dx^2} = \frac{(n-1)Tan\,\alpha_E}{x_E^{n-1}}\,x^{n-2} \quad\text{...} \quad [1.8.3]$$

Uniform Twist of Rifling

In the case of uniform twist of rifling the exponent $n = 1$

Equation [1.8.1] becomes:

$$y = Kx$$

Equation [1.8.2] becomes:

$$\frac{dy}{dx} = K = Tan\,\alpha_E$$

And:

$$\frac{d^2y}{dx^2} = 0$$

Similarly $\frac{dy}{dx}$ and $\frac{d^2y}{dx^2}$ can be evaluated for any value assigned to n.

Estimation of Rifling Torque

Now with the help of Equations [1.8.2] and [1.8.3] and the expression for force on the driving band vide Equation [1.7.1] Reference 1 it is possible to estimate the rifling torque on the driving band within the accuracy of the assumptions made in deriving Equation [1.7.1] Reference 1.

Force on the driving band is:

$$F_{DB} = \left(\frac{2k}{D}\right)^2 \left[\frac{dy}{dx}P + m_p v_p^2 \frac{d^2 y}{dx^2}\right] \quad \text{..................................... Equation [1.7.1] Reference 1}$$

k: radius of gyration of projectile
D: calibre
P: propellant gas force on projectile base
m_p: projectile mass
v_p: projectile velocity

Hence, torque on the driving band is:

$$T_{DB} = \{\frac{2k}{D}\}^2 \left[\frac{dy}{dx} + m_p v_p^2 \frac{d^2 y}{dx^2}\right]\frac{D}{2} \quad \text{.. [1.8.4]}$$

Computation of the torque on the driving band for different rifling curves is illustrated with the help of an example.

Design Exercise 1.8.1

Compute the torque on the driving band of the 76 mm HE projectile for rifling curves of uniform twist, and of increasing twist of exponent values n = 1.0, 1.25, 1.5, 1.75 and 2, given the following information:

Mass of the projectile is 6.2 kg, the $\frac{2k}{D}$ ratio is 0.75 for the given HE shell. The angle of rifling at exit of the projectile is 5°. Other necessary information is available in the worksheet given in Fig 1.8.1:

Solution

The problem is solved as follows:

Values of $\dfrac{dy}{dx}$ and $\dfrac{d^2y}{dx^2}$ are computed for different values of exponent n, vide Equations [1.8.2] and [1.8.3] for values of shot displacement obtained from the internal ballistics worksheet.

The propellant gas pressure and projectile velocity corresponding to the same shot displacement is also obtained from the internal ballistics spread sheet.

The propellant gas pressure multiplied by the area of the projectile base gives the gas force acting on the base of the projectile. This value along with the projectile velocity and mass of projectile are substituted in Equation [1.8.4] to estimate the rifling torque from shot start to muzzle.

The computer programme for computation of the rifling torque is given below and the results shown graphically in Fig 1.8.2:

Symbols used in the worksheet:

ve: muzzle velocity
mc: charge mass
mp: projectile mass
l: total shot travel
vp: projectile velocity
s: shot travel
p: propellant gas pressure
pm: maximum propellant gas pressure
subscript m indicates point of maximum pressure

Internal ballistics worksheet for computation of rifling torque					
ve	680	vm	258.4	l	2.587
mc	1.08	tm	0.002480476	sigmanetap	0.0875
pm	226933980	sm	0.2263625	mp	6.2
lamda	s	psi	phi	p	vp
0.00	0.000	0.000	0.000	0.00E+00	0.000
0.25	0.057	0.741	0.392	1.68E+08	101.293
0.50	0.113	0.912	0.635	2.07E+08	164.084
0.75	0.170	0.980	0.834	2.22E+08	215.506
1.00	0.226	1.000	1.000	2.27E+08	258.400
1.25	0.283	0.989	1.140	2.24E+08	294.576
1.50	0.340	0.965	1.262	2.19E+08	326.101
1.75	0.396	0.932	1.366	2.12E+08	352.974
2.00	0.453	0.898	1.468	2.04E+08	379.331
2.50	0.566	0.823	1.632	1.87E+08	421.709
3.00	0.679	0.747	1.763	1.70E+08	455.559
3.50	0.792	0.675	1.875	1.53E+08	484.500
4.00	0.905	0.604	1.983	1.37E+08	512.407
4.50	1.019	0.546	2.068	1.24E+08	534.371
5.00	1.132	0.495	2.140	1.12E+08	552.976
6.00	1.358	0.403	2.269	9.15E+07	586.310
7.00	1.585	0.338	2.363	7.67E+07	610.599
8.00	1.811	0.284	2.445	6.44E+07	631.788
9.00	2.037	0.248	2.509	5.63E+07	648.326
10.00	2.264	0.220	2.566	4.99E+07	663.054
11.00	2.490	0.199	2.615	4.52E+07	675.716
11.42	2.585	0.191	2.633	4.34E+07	680.367

Fig 1.8.1: Internal ballistics worksheet for computation of rifling torque

Matlab Programme 1.8.1

```
% programme to compute rifling torque
% file name: rifling torque.txt

twok_by_D=.73 %HE shell
mp=6.2 % mass of shell kg
tan_alfa_e=.0872664
s_e=2.585
D=.076

rng='e6..e27'
rng1='b6..b27'
rng2='f6..f27'
p=wk1read('a:rifle',3,4,rng)
P=p.*.076.^2.*pi./4
s=wk1read('a:rifle',3,1,rng1)
v=wk1read('a:rifle',3,5,rng2)

n1=1
dybydx1=tan_alfa_e/s_e.^(n1-1).*s.^(n1-1)
d2ybydx21=(tan_alfa_e.*(n1-1)./s_e.^(n1-1)).*s.^(n1-2)
T1=(twok_by_D).^2.*(dybydx1.*P+mp.*v.^2.*d2ybydx21).*D./2

n2=1.25
dybydx2=tan_alfa_e/s_e.^(n2-1).*s.^(n2-1)
d2ybydx22=tan_alfa_e.*(n2-1)./s_e.^(n2-1).*s.^(n2-2)
T2=(twok_by_D).^2.*(dybydx2.*P+mp.*v.^2.*d2ybydx22).*D./2

n3=1.5
dybydx3=tan_alfa_e/s_e.^(n3-1).*s.^(n3-1)
d2ybydx23=tan_alfa_e.*(n3-1)./s_e.^(n3-1).*s.^(n3-2)
T3=(twok_by_D).^2.*(dybydx3.*P+mp.*v.^2.*d2ybydx23).*D./2

n4=1.75
dybydx4=tan_alfa_e/s_e.^(n4-1).*s.^(n4-1)
d2ybydx24=tan_alfa_e.*(n4-1)./s_e.^(n4-1).*s.^(n4-2)
T4=(twok_by_D).^2.*(dybydx4.*P+mp.*v.^2.*d2ybydx24).*D./2
```

```
n5=2
dybydx5=tan_alfa_e/s_e.^(n5-1).*s.^(n5-1)
d2ybydx25=tan_alfa_e.*(n5-1)./s_e.^(n5-1).*s.^(n5-2)
T5=(twok_by_D).^2.*(dybydx5.*P+mp.*v.^2.*d2ybydx25).*D./2

plot(s,T1,s,T2,s,T3,s,T4,s,T5)
title('Force on driving band for different rifling curves')
xlabel('shot travel');ylabel('Force on driving band')
gtext('n=1.0');gtext('n=1.25');gtext('n=1.5');gtext('n=1.75');gtext('n=2.0')
```

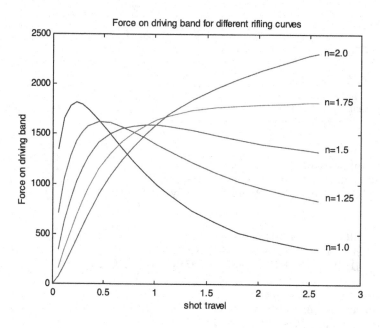

Fig 1.8.2: Plot showing rifling torque for different rifling curves

Selection of Rifling Curve

From Fig 1.8.2 it emerges that a uniform twist of rifling gives rise to high torques towards the beginning of projectile travel. The imposition of torque on the driving band in the case of increasing twist is gradual as compared to that of uniform twist. However the magnitude of rifling torque on the driving band, at exit, is the less in the case of uniform twist. Hence in the case of increasing twist, the stripping of the driving band is less likely to occur, wear on the lands is also less. With uniform twist, the driving band is engraved exactly according to the rifling profile during its axial displacement within the forcing cone. The full depth of the grooves is realized at the commencement of rifling, after which there is no further displacement of the driving band material. Once engraved, resistance to the motion of the projectile becomes purely frictional. With increasing twist of rifling, the angle of rifling changes continuously and the engraving process is continuous until the driving band clears the muzzle. This results in greater stresses at the forward portion of the band, increased friction, abrasion and resistance to the passage of the projectile for the entire duration of its passage up the bore. Of importance also is that the manufacturing process of increasing twist is far more intricate and costly as compared to that of uniform twist.

The final choice of the rifling curve has to be a compromise between influences such as high initial and low exit torque, cost and ease of manufacture and choice of driving band material.

With reference to Fig 1.8.2, the curve with an exponent value of around 1.5 appears to offer the best technical solution. However, simplicity and the economics of manufacturing the selected rifling curve, as well as the effect of the selected rifling curve on barrel life also call for consideration.

Rifling Profile

Rifling profile defines the number of grooves, the depth of grooves and the cross sectional shape of the grooves and lands. Based on the minimum width of the land from material strength considerations, larger calibres can have a greater number of lands and grooves resulting in reduced resistance to engraving. But guns with lesser number of lands have a longer accuracy life. The shape of the land groove combination influences the direction in which the force is exerted on the driving band. This should ideally be tangential. To meet this requirement, the driving edge of the land should be radial and its face flat. Rounding off of the groove root reduces the magnitude of stress concentration and renders the profile uncomplicated to cut. The shape of the following edge of the land is not critical so long as it does not induce weakness in the land. Generally accepted profiles consist of both walls cut to a taper of around 10°. This prevents the lands from being stripped during manufacture. The corners at the groove root are rounded to reduce stress concentration and sharp edges of the lands chamfered to reduce the effect of erosion.

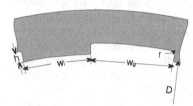

Fig 1.8.3: Parameters of rifling profile

Number of Grooves

From practical experience the number of grooves (or lands for that matter) calibre relation ship has been approximately established as 8 grooves per 25.4 mm of calibre. For instance the total number of grooves required in a 105 mm gun is:

$$G = 8\frac{105}{25.4} = 33$$

The calculated and the actual number of grooves obtained from statistical sources are compared in tabular form in Fig 1.8.5 and graphically in Fig 1.8.6.

Ratio of Land to Groove Width

Narrow grooves result in narrow ribs being cut into the driving band. The driving band being of comparatively softer material than the barrel, such a situation renders the driving band prone to shearing. In such event, the rifling will not rotate the projectile properly. It is hence accepted practice to have the rifling grooves wider than the lands. Based on practical experience, different land to groove width ratios of 2:3 to 1:2 have been satisfactorily used in practice.

Groove and Land Width

Accepting a land to groove width of 2:3 and the number of grooves to calibre relationship as 8 grooves per 25.4 mm of calibre, the groove and land width become constant as follows:

Length of circumference for 25.4 mm of calibre: $\pi 25.4 = 79.796$ mm

Width of one groove + width of one land = 9.974 mm

Width of one groove:

$$w_g \frac{9.974}{5} 3 = 5.9844 \ \text{mm}$$

Width of one land:

$$w_l : \frac{9.974}{5} 2 = 3.9898 \ \text{mm}$$

Practically, when using this relation, some adjustment may be required to fit the total width of grooves and lands into the bore circumference.

Number of Grooves from Groove and Land Width

Another approach to fix the number of grooves is based on the calculated width of grooves and lands.

From the point of view of resistance to engraving, the width of the lands should be the least possible. But the width of the land is also related to the maximum thrust on the land and the shear strength of the material of the gun barrel.

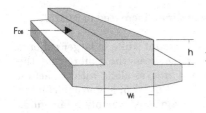

Fig 1.8.4: Leading dimensions of a land

With reference to Fig 1.8.6 and considering that the maximum pressure on the land is around 0.8 G Pa and the shear stress of barrel steel is in the region of 0.32 G Pa, the force on unit length of the driving face is: $F_{DB} = 0.8h$ G N

Hence width of the land in terms of its height per unit length of land is:

$$w_l = \frac{0.8h}{0.32} = 2.5h$$

The width of grooves which is calculated similarly making use of the shear strength of copper is:

$$w_g = 5.0h$$

It follows that the groove to land width obtained from this approach is 2:1. Knowing the calibre, the width of one groove and land pair can be calculated, hence the number of grooves.

Depth of Grooves/Height of Lands

Deep grooves tend to diminish the strength of the barrel, whereas, shallow grooves do not weaken the barrel but get worn out, markedly on the driving edge, after prolonged usage resulting in loss of accuracy. Based on this reasoning, an accepted guideline for fixing the depth of grooves is:

$$h = 0.01D$$

D: calibre of the weapon

To cater for the effect of wear the groove height is sometimes increased to $0.0125D$.

Smaller calibres have grooves depths in excess of this amount while in larger calibres the groove depth may be slightly less. The relation between the calibre and the calculated depth of grooves, width of lands and width of grooves along with the actual data obtained from practice is shown in tabular form in Fig 1.8.5, and graphically in Fig 1.8.6. Clearly, the rules given in the preceding paragraphs are only a fair guide and no more.

Calibre D	No of grooves calculated	No of grooves actual	Depth of groove calculated h	Depth of groove actual	Groove width calculated wg	Groove width actual	Land width calculated wl	Land width actual	Radius at groove root r
20.0	6	9	0.200	0.427	5.98	5.2070	3.990	5.2070	0.254
30.0	9	16	0.300	0.494	5.98	3.4290	3.990	2.4892	0.254
40.0	13	16	0.400	0.599	5.98	5.5880	3.990	2.2606	0.610
57.0	18	24	0.570	0.520	5.98	4.4094	3.990	3.0480	0.508
75.0	24	28	0.750	0.778	5.98	4.7396	3.990	3.6678	0.381
76.2	24	28	0.762	1.067	5.98	4.7396	3.990	3.8100	0.381
90.0	28	32	0.900	1.063	5.98	5.0241	3.990	3.8100	0.381
105.0	33	36	1.050	1.196	5.98	5.3543	3.990	3.8100	0.381
152.4	48	48	1.524	1.346	5.98	6.3322	3.990	3.6424	0.508
155.0	49	48	1.550	1.316	5.98	6.3322	3.990	3.8100	0.508

Fig 1.8.5: Calculated and actual data of rifling profiles

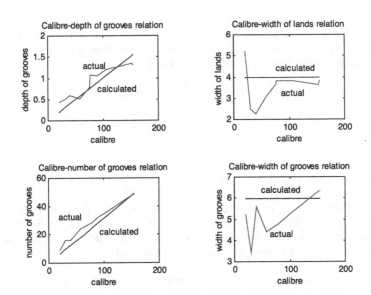

Fig 1.8.6: Calibre versus, depth of grooves, width of lands, number of grooves and width of grooves calculated and actual relations

Effect of the Shape of the Driving Edge of the Groove

The shape of the driving edge has considerable influence on the thrust producing rotation. Different profiles have different radii of the groove root curve, which may also be different for the driving edge and the following edge of the land.

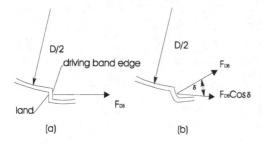

Fig 1.8.6: Effect of shape of driving edge of groove on torque on driving band. (a) Flat driving edge (b) Curved driving edge

From the figure it is seen that if the surface of the driving edge of the groove lies on a line joining its root to the bore axis, then the thrust on the driving band will act tangentially to the circumference of the driving band.

With reference to Fig 1.8.6 (a), the torque producing rotation is:

$$T_{DB} = \frac{D}{2} F_{DB}$$

D: calibre
F_{DB}: Force on the driving band

Any change to the shape of the driving edge of the groove will cause the thrust on the driving band to act in a direction, which is other than tangential to the circumference of the driving band. The thrust may be resolved into its normal component producing rotational torque and its component acting radially inward.

The torque producing rotation in this case depicted in Fig 1.8.6 (b) is:

$$T_{DB} = \frac{D}{2} F_{DB} Cos\delta$$

δ : angle between the tangent to the circumference and the direction of the thrust on the driving band at the point of contact

Obviously the torque producing rotation is less for groove shapes other than parallel sided.

Polygroove Plain Section

This is one accepted form of rifling groove. Here the grooves are parallel sided with a radius at the groove root, which is usually equal to the groove height. The corners of the lands are not curved or chamfered.

Selection of Rifling Profile

For a new barrel it is the accepted practice to select an already developed standard rifling profile for the calibre in question. A standard rifling profile is a good start to the rifling profile design. Another option open to the designer is to choose a standard profile and go on to modify the driving band to suit the chosen profile.

Cutting of Rifling

Rifling grooves may be incised on the bore surface by cutting or by cold forging. Cutting is the preferred method for pre stressed gun barrels while cold forging is suitable for gun barrels of monobloc construction.

1.9 Chamber Design

Basis of Chamber Design

Chamber design is based on internal ballistics considerations and ammunition dimensions. If the ammunition is being designed from scratch along with the barrel, then a joint approach by the ballistician, ammunition designer and the gun designer is advocated. If existing ammunition is to be used, the chamber is designed to provide the stipulated combustion space, at the same time catering for ease of loading and extraction. Chamber design involves the determination of the following three chamber dimensions, i.e. chamber mouth diameter, chamber length and length of the forcing cone.

Definitions Associated with Chamber Design

Chamber Face: This is the breech face of the barrel into which the chamber is cut. In weapons of larger calibre, the cartridge case flange rests on this face or inside a suitable counter boring on the face. In both cases grooves are cut on either side to accommodate the extractor lips.

Chamber Mouth: This is the circular opening of the chamber at the chamber face.

Chamber Mouth Diameter: This is the diameter of the chamber mouth. It is kept as small as possible to reduce breech end loading. Breech end loading is the product of the gas pressure and inner area of the cartridge case base. The chamber mouth diameter is also kept small so as to have the smallest external diameter of the barrel over the chamber as this is the heaviest portion of the barrel and influences the overall weight of the equipment.

Chamber Volume: In the case of fixed ammunition, the chamber volume is the volume in the empty cartridge case with projectile fixed. In the case of separate ammunition it is the chamber volume up to the circle of contact of the driving band with the chamber wall, less the volume of the base of the projectile extending backwards into the chamber.

Chamber Length: This is the distance from the chamber face to the base of the projectile when loaded. A long and thin chamber is desirable from the lightness point of view. In turret-mounted guns there is a limitation of space within the fighting compartment, which also has to accommodate the movement of the barrel in recoil. In such circumstances, the chamber perforce has to be short and wide.

Chamber Slope: Chamber slope is introduced to support easy extraction of the fired cartridge case. Chamber slopes ranging from 20' to 2° are in use. Both chamber and cartridge case must necessarily have identical taper.

Centering Cylinder: This is the cylindrical portion between the body of the chamber and the forcing cone in which the projectile longer axis and the bore axis are brought into alignment. The diameter of this cylinder conforms to that of the driving band plus some nominal clearance.

Forcing Cone: This is the frustum of a cone interposed between the centering cylinder and the origin of rifling. The slope of the forcing cone or the semi cone angle dictates the impact velocity of the driving band when it contacts the rifling. Semi cone angles from 1° to 9° are in use. A smaller angle permits longer projectile travel before impact resulting in higher impact velocities. A steeper slope lessens the impact velocity but is accompanied by greater erosion. In the case of separate ammunition the risk of projectile slip back is greater. Fig 1.9.1 refers.

The forcing cone length is obtained from the height of the driving band and the selected forcing cone semi angle.

Height of the driving band is calculated from the approximate expression:

$$h = \left(r_c + g + .254 \right) - r_c \text{ mm} \quad\text{...} [1.9.1]$$

Length of the forcing cone is hence given by:

$$l_{fc} = \frac{h}{Tan\,\theta} \quad\text{...} [1.9.2]$$

r_c: calibre radius
g: depth of groove usually taken as 1% of calibre
θ : forcing cone semi angle

Chamber Design Considerations

Chamber-Cartridge Case Interaction

On firing, the walls of the cartridge case, which are thin towards the lip, expand elastically under the action of the propellant gas pressure and as the pressure increases beyond that corresponding to the elastic limit the deformation becomes plastic. The expansion continues until the outer surface of the case walls comes in close contact with the inner surface of the chamber. The thickness of the case wall is insufficient for the material of the case to withstand the pressure of the propellant gases independently. As the pressure increases further, the chamber wall absorbs the force of the propellant gas acting on the inner surface of the cartridge case wall by expanding elastically. The maximum pressure of the propellant gas, the thickness of the walls and the properties of the material of the chamber wall dictate the magnitude of expansion of the chamber wall. After projectile exit, the internal pressure in the cartridge case drops. The chamber, which is designed to expand well within its elastic limit, returns to its original dimensions. The cartridge case should now contract to a greater extent than the chamber walls in order to be ready for extraction. The cartridge case material must therefore have a high resilience to fulfill the requirement of extraction.

The initial radial clearance between the case outer radius and the chamber inner radius is critical. If this clearance is excessive, the case will rupture during expansion, if small; it results in difficulty during loading.

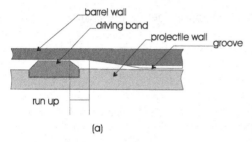

(a)

Fig 1.9.1: Effect of forcing cone slope (fixed ammunition). (a) Shallow slope. (b) Steep slope

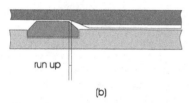

(b)

Mechanics of Chamber-Cartridge Case Recovery

With reference to Fig 1.9.2, the total radial deformation of the case is the sum of the initial clearance plus the elastic radial deformation of the chamber. On release of the pressure, the chamber will return to its original dimensions, but the case, which has suffered permanent set, will not return to its original dimensions. In case the permanent set is less then the initial clearance, a clearance will exist between the case and the chamber wall after firing; otherwise interference occurs, rendering extraction difficult. It follows that for extraction of the fired case two conditions should be met. Firstly the recovery of the cartridge case must be greater than that of the chamber and secondly the initial clearance must exceed the permanent set of the case material.

The radial elastic strain of the chamber wall, which is a thick cylinder, can be calculated from:

$$\varepsilon_c = \frac{p_m}{E_c}\left(\frac{T_c^2 + 1}{T_c^2 - 1} + \mu_c\right)$$ .... Equation [1.2.16] Reference 1

p_m: maximum propellant gas pressure
E_c: Youngs Modulus of chamber material
T_c: wall ratio of chamber
μ_c : Poisson's Ratio of chamber material

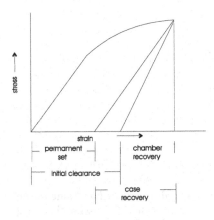

Fig 1.9.2: Chamber-cartridge case recovery

Treated as a thin cylinder enclosed at both ends, at yield, the axial stress of the cartridge case is:

$$\sigma_l = \frac{p_{yp}r}{2t}$$ .... [1.9.3]

83

p_{yp}: yield stress of the case material
r: internal diameter of the case
t: case thickness

The hoop stress of the case is given by:

$$\sigma_t = \frac{p_{yp} r}{t} \quad\text{...}\quad [1.9.4]$$

The elastic resilience of the cartridge case is given by:

$$\varepsilon_{cc} = \frac{p_{yp} r}{2t E_{cc}} \left(2 - \mu_{cc}\right) \quad\text{..}\quad [1.9.5]$$

ε_{cc} : radial displacement of cartridge case
p_{yp} : pressure at yield point of case material
E_{cc}: Young's Modulus of case material
μ_{cc} : Poisson's Ratio of case material

For easy extraction, as earlier mentioned, the condition is:

$$\varepsilon_{cc} > \varepsilon_c$$

From Equations [1.9.5] and [1.2.16] Reference 1:

$$\frac{p_{yp}}{2tE}\left(2 - \mu_{cc}\right) > \frac{p_m}{E_c}\left(\frac{T_c^2 + 1}{T_c^2 - 1} + \mu_c\right)$$

For a given peak pressure and specified chamber material, the right hand side of the above equation can be decreased by increasing the wall ratio of the chamber. Fig 1.9.3 shows the relationship between radial elongation and wall ratio.

Chamber Wall-Cartridge Case Clearance

Practically the initial chamber wall-cartridge case clearance is taken as 2% of the bore radius.

Longitudinal Clearance

Simultaneously, the cartridge case is stressed longitudinally. The gas pressure acting on the base of the cartridge case bears against the surface of the breech block. The cartridge head clearance is taken up, by longitudinal expansion of the case material, whilst the lip of the case is held tightly against the chamber walls, resulting in longitudinal stress in the material of the cartridge case. If this stress is excessive the case will rupture transversely towards its base, leaving some part of the case still inside the chamber during extraction. This condition is obviated during the design stage of the chamber by providing a small taper. A small taper serves in reducing friction between the case and chamber wall, improving the operations of feed and extraction. Also during extraction, a small movement of the cartridge case in the axial direction is sufficient to break contact between the chamber walls and the cartridge case outer surface, affording easy extraction. The stipulated combustion volume of the case limits the magnitude of this taper.

Axial clearance between the case and chamber must also be considered. If this clearance is excessive, the case is likely to rupture transversely. During firing the lip expands to hold the chamber wall while simultaneously the gas force acts on the inside of the cartridge case base. In chamber design usually an empirically established longitudinal clearance of 2.5×10^{-4} m is incorporated.

Flange Dimensions

The following relations give the flange dimensions:

Flange radius: 0.5375 x chamber mouth diameter

Flange thickness

Chamber mouth diameter mm	Flange thickness mm
115	4.5
115-180	6.0
>180	7

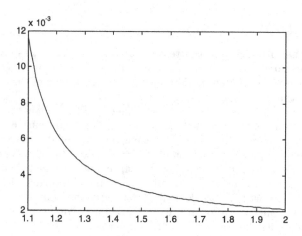

Fig 1.9.3: Chamber wall ratio-radial elastic displacement curve

1.10 Preliminary Design of a Gun Barrel Chamber

Design Exercise 1.10.1

Design of a Chamber with Known Ammunition Dimensions

The chamber for a 76 mm gun is to be designed for existing ammunition given the following ballistics and other relevant information:

Calibre radius: 0.038 m
Combustion volume: 0.0246 m^3
Chamber mouth diameter is to be kept a minimum
Tolerance to be maintained: 0.06 mm
Forcing cone slope: 2°

Case Thickness

For preliminary design purposes, the cartridge case thicknesses may be taken from the table in Fig 1.10.1 ahead:

Calibre mm	Thickness one calibre from base mm	Lip thickness mm
< 80	1.9	1.14
>=80	2.286	1.524

Fig 1.10.1: Calibre-cartridge case thickness rough estimation guide

Chamber Design Parameters

As mentioned in the previous section, the three dimensional parameters of the chamber to be established are the chamber mouth diameter, chamber length and length of the forcing cone.

Determination of Chamber Mouth Diameter

The chamber mouth diameter is determined starting from the known calibre radius according to the following procedure, starting at the lip end of the cartridge case:

High outer radius of the cartridge case. To the bore radius is added the case thickness obtained from Fig 1.10.1: and the given tolerance. This gives the high outer radius of the cartridge case.

Low inner radius of chamber. The chamber-case clearance is added to the high outer radius of the cartridge case. This gives the low inner radius of the chamber.

Length of chamber. The chamber length is now computed making use of the given combustion volume and the area of cross section at the cartridge case lip.

Chamber mouth diameter. The chamber mouth diameter is calculated by adding the increase in radius, due to the selected taper, to the chamber radius at the lip.

Inner Radius of Cartridge Case One Calibre from Bottom

Chamber inner radius one calibre from bottom. This is obtained by reducing from the chamber mouth radius the amount due to the taper over the length of one calibre.

High outer radius of case. This is obtained by subtracting the tolerance and the chamber-case clearance from the chamber inner radius.

Internal radius of case. The internal radius of the case is got by reducing the case thickness, obtained from Fig 1.10.1, from the high outer radius of the case.

Combustion volume. The combustion volume is now found by summing of the volumes up to one calibre from base and from this point to the lip. By comparison with the specified combustion volume it is established whether the computed chamber length is correct, in excess or less than required. A correction is now made to the initial computed chamber length.

The chamber mouth diameter and the chamber length have been established in the worksheets of Fig 1.10.2: and Fig 1.10.3:

	A	B
1	Chamber Design	
2	Data	
3	Calibre radius	0.038
4	Combustion volume	0.00246
5	Taper	0.01
6	Dimension at lip	
7	Case thickness	0.00152
8	Outer radius of case	=B3+B7
9	Tolerance	0.00006
10	High outer radius of case	=B8+B9
11	Chamber- case clearance	=0.002*B3
12	Low radius of chamber	=B10+B11
13	Length of chamber	=B4/(PI()*B12^2)
14	Chamber mouth radius	=B3+B13*B5
15	Radius inside case one calibre from bottom	
16	Chamber radius	=B14-(B5*0.076)
17	Tolerance	=B9
18	Low radius of chamber	=B16-B17
19	Clearance chamber to case	=B11
20	High outer radius of case	=B18-B19
21	Tolerance	=B9
22	Low outer radius of case	=B20-B21
23	Case thickness	0.00229
24	Internal radius of case	=B22-B23
25	Comparison of computed & actual chamber volume	
26	Volume one calibre from base	=B24^2*PI()*0.076
27	Volume from base to lip	=((B24-B3)/2+B3)^2*PI()*(B13-B3*2)
28	Computed combustion volume	=B26+B27
29	Excess/deficient combustion volume	=B4-B28
30	Computed chamber length	=B28/(B24^2*PI())
31	Excess computed chamber length	=B13-B30
32	Adjusted chamber mouth radius	=B14+B31*B5

Fig 1.10.2: Worksheet for chamber design of 76 mm gun (data & formulae)

Chamber Design	
Data	
Calibre radius	0.03800
Combustion volume	0.00246
Taper	0.01000
Dimension at lip	
Case thickness	0.00152
Outer radius of case	0.03952
Tolerance	0.00006
High outer radius of case	0.03958
Chamber- case clearance	0.00008
Low radius of chamber	0.03966
Length of chamber	0.49793
Chamber mouth radius	0.04298
Radius inside case one calibre from bottom	
Chamber radius	0.04222
Tolerance	0.00006
Low radius of chamber	0.04216
Clearance chamber to case	0.00008
High outer radius of case	0.04208
Tolerance	0.00006
Low outer radius of case	0.04202
Case thickness	0.00229
Internal radius of case	0.03973
Comparison of computed & actual chamber volume	
Volume one calibre from base	0.00038
Volume from base to lip	0.00200
Computed combustion volume	0.00238
Excess/deficient combustion volume	0.00008
Computed chamber length	0.47972
Excess computed chamber length	0.01821
Adjusted chamber mouth radius	0.04316

Fig 1.10.3: Worksheet for chamber design of 76mm gun of (data & results)

Estimation of Forcing Cone Length

From Equation [1.9.1], the height of the driving band is:

$h = (38 + .01x.38 + .254) - 38 = 38.634$ mm

The forcing cone length using Equation [1.9.2] is:

$$l_{fc} = \frac{0.634}{Tan1^o} = 36.32 \text{ mm}$$

1.11 Vibration and Accuracy of Gun Barrels

Vibration Effects on Gun Barrels

Gun barrels are subject to two effects due to vibrations. Firstly vibrations set up by the passage of the projectile induce vibrations in the radial direction due to the expansion of the barrel when the projectile enters a section of the barrel and subsequent contraction when the projectile leaves the section. This effect is particularly harmful at the muzzle end where the barrels are thinner than the rest of the barrel and also under the influence of open-end effects. This problem may be overcome by suitably strengthening the barrel wall at the muzzle. The second more immediately undesirable effect is that on the accuracy of the weapon.

Vibrations of Gun Barrels and Effect on Accuracy

Trial firings by gun designers have established a relationship between the accuracy of target engagement and the ratio of the frequency of vibration of the gun barrel itself, also called its natural frequency, to the rate of fire of the weapon. This ratio is called the Frequency Ratio and is given by:

$$F_r = \frac{f_n}{f_{rf}} \quad \text{..} \quad [1.11.1]$$

f_n: natural frequency of the gun barrel c/s
f_{rf}: frequency of fire rounds/s

The relation between accuracy and Frequency Ratio is depicted in Fig: 1.11.1

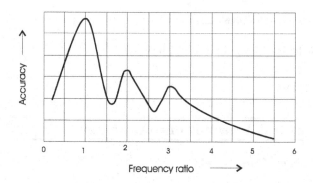

Fig 1.11.1: Plot of accuracy-Frequency Ratio relationship

As evident from Fig 1.11.1, ideally a gun barrel should be designed for a Frequency Ratio greater than 4.0. The natural frequency of a gun barrel depends on its stiffness which is a function of wall thickness and rigidity of the material. In case it is not possible to increase the wall thickness due to other, more pressing, considerations the barrel is designed so that the magnitude of the Frequency Ratio falls at the closest of the troughs of the accuracy-Frequency Ratio curve.

In particular cases it may be necessary to adjust the round to round time interval in order to minimize the negative accuracy effect of vibration.

Estimation of Natural Frequency of a Gun Barrel

In order to estimate the natural frequency of a gun barrel it is necessary firstly to estimate its deflection. A gun barrel is characteristically a barrel of homogeneous material but varying cross section, therefore varying weight distribution. Estimation of the deflection of a gun barrel possessing these characteristics may be conveniently done by application of the area-moment method.

Area Moment Method for Estimation of Deflection of a Gun Barrel

From the pure bending of a uniform beam we have the relationship:

$$\frac{1}{r} = \frac{M}{EI} \quad \text{.. [1.11.2]}$$

r: radius of curvature of the deflection curve
M: bending moment at the cross section under consideration
I: moment of inertia of the cross section at the point under consideration about the neutral axis

Considering the gun barrel as a cantilever it is now necessary to determine the maximum displacement of the barrel from its original position.

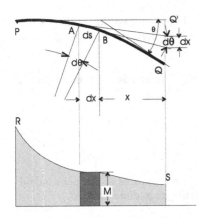

Fig 1.11.2: Deflection of a beam by the Area-Moment method

With reference to Fig 1.11.2, PQ represents a part of a deflection curve of a gun barrel and RS the corresponding part of the bending moment diagram.

Taking an angle $d\theta$ subtended at the centre of the deflection curve by an element of the barrel:

$ds = rd\theta$

Applying Equation [1.11.2]:

$$d\theta = \frac{M}{EI} ds$$

Since in the case of a gun barrel the deflection is very small, i.e. of the order 10^{-3}, it can be approximated that:

$$ds = dx$$

Hence:

$$d\theta = \frac{M}{EI} dx$$

Graphically this means that the elemental angle between two tangents to the deflection curve is equal to the elemental area Mdx of the bending moment diagram divided by EI.

It follows that the angle between the tangents at the extremities can be obtained from:

$$\theta = \int_{P}^{Q} \frac{1}{EI} Mdx$$

In other words, the angle between the tangents at the extremities of the bent barrel equals the area of the corresponding bending moment diagram divided by EI.

Consider the vertical displacement of the end Q due to the bending of an element AB of the barrel. The displacement caused by the bending of this element is:

$$xd\theta = x\frac{1}{EI} Mdx$$

Graphically this is the moment of the bending moment Mdx about the vertical through P, divided by EI.

Symbolically the total vertical displacement of P is given by:

$$y = \int \frac{1}{EI} Mx dx$$

In other words, the vertical displacement or the deflection of the extremity Q is equal to the moment about the vertical through Q of the area of the bending moment of the entire beam, divided by EI.

Deflection of a Gun Barrel considered a Cantilever with Uniform Load

With reference to Fig: 1.11.3 the bending moment at any section C-C distant x from the fixed end is:

$$M = w(l-x)\frac{l-x}{2}$$

Or:

$$M = \frac{w}{2}(l-x)^2$$

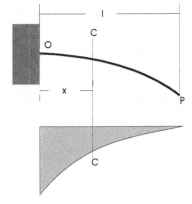

Fig 1.11.3: Gun barrel as a cantilever with uniform load distribution

The deflection at a distance l from the fixed end is the moment of the area shown shaded about the vertical through P, the moment arm being the horizontal distance from the centroid of the shaded area to the vertical through P, divided by EI.

Determination of the Natural Frequency

This method is based on the equality of the maximum kinetic energy of the barrel at the mean position and the maximum potential or strain energy at the extreme position or the amplitude position.

Assuming simple harmonic motion of the barrel, displacement of the centre of the end of the barrel:

$$y = Y \sin \omega t \quad \dots\dots\dots\dots\dots\dots\dots\dots\dots\dots\dots\dots\dots\dots\dots\dots\dots\dots\dots\dots\dots\dots\dots\dots \text{[1.11.3]}$$

y: vertical displacement of the point at time t
ω: angular speed radians/s
Y: amplitude

Differentiating Equation [1.11.3]:

$$\frac{dy}{dt} = \omega Y \cos \omega t$$

Maximum velocity at the mean position is:

$$\frac{dy}{dt} = \omega Y$$

Hence the maximum kinetic energy is:

$$E_{ke} = \frac{1}{2} m_b \omega^2 Y^2$$

m_b: mass of the barrel

Differentiating Equation [1.11.3] again:

$$\frac{d^2 y}{dt^2} = -\omega^2 Y \sin \omega t$$

It follows that the maximum acceleration at $\omega t = \dfrac{\pi}{2}$ or multiples of it is:

$$\frac{d^2 y}{dt^2} = -\omega^2 Y \quad \text{.. [1.11.4]}$$

The negative sign here implies sense towards the mean position. The maximum potential energy at the extreme position is given by:

$$E_{pe} = \frac{0 + sY}{2} Y$$

s: stiffness of the barrel

Or:

$$E_{pe} = \frac{1}{2} s Y^2$$

Equating the maximum kinetic energy and maximum potential energies:

$$\omega^2 = \frac{s}{m}$$

The natural frequency is given by:

$$f_n = \frac{\omega}{2\pi}$$

Or:

$$f_n = \frac{1}{2\pi} \sqrt{\frac{s}{m}} \quad \text{.. [1.11.5]}$$

Natural Frequency of Vibration of a Gun Barrel Subject to Lump Loading

A gun barrel may be considered a cantilever with uniform loading over the part of its length wherein the barrel is of uniform cross section and with varying load over the part where it tapers. It is convenient in such instances to consider the barrel as a beam with lumped loads along its length. One method, which can be readily applied, for determination of the natural frequency of a gun barrel is discussed below.

Rayleighs (Energy Method) for Determination of the Natural Frequency of a Gun Barrel.

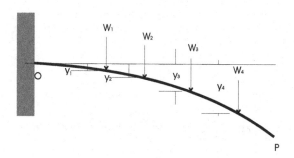

Fig 1.11.4: Deflection of a gun barrel by the lumped loading method

With reference to Fig 1.11.4, $Y_1, Y_2, \ldots \ldots Y_n$ are the net maximum deflections of a barrel under n lumped loads $W_1, W_2, \ldots \ldots W_n$ respectively.

The maximum potential energy is hence:

$$E_{pe} = \frac{1}{2}W_1Y_1 + \frac{1}{2}W_2Y_2 \ldots \ldots + \frac{1}{2}W_nY_n$$

Or:

$$E_{pe} = \frac{1}{2}\sum_1^n W_nY_n$$

The maximum kinetic energy is:

$$E_{ke} = \frac{1}{2}\frac{W_1}{g}\left(\omega Y_1\right)^2 + \frac{1}{2}\frac{W_2}{g}\left(\omega Y_2\right)^2 \ldots\ldots\ldots\ldots + \frac{1}{2}\frac{W_n}{g}\left(\omega Y_n\right)^2$$

Or:

$$E_{ke} = \frac{1}{2}\frac{\omega}{g}\sum_1^n W_n Y_n^2$$

Equating the maximum kinetic energy to the maximum potential energy:

$$\omega^2 = g\frac{\sum_1^n W_n Y_n}{\sum_1^n W_n Y_n^2}$$

Finally:

$$\omega = \sqrt{g\frac{\sum_1^n W_n Y_n}{\sum_1^n W_n Y_n^2}} \quad\ldots\ldots\ldots\ldots\ldots\ldots\ldots\ldots\ldots\ldots\ldots\ldots\ldots\ldots\ldots\ldots\ldots\ldots\ldots\ldots\ldots\ldots\ldots \text{[1.11.6]}$$

ω : circular frequency of vibration radians/s

The natural frequency of vibration of the barrel is:

$$f_n = \frac{1}{2\pi}\sqrt{g\frac{\sum_1^n W_n Y_n}{\sum_1^n W_n Y_n^2}} \quad c/s \quad\ldots\ldots\ldots\ldots\ldots\ldots\ldots\ldots\ldots\ldots\ldots\ldots\ldots\ldots\ldots\ldots \text{[1.11.7]}$$

1.12 Vibration Characteristic of a Gun Barrel

Design Exercise 1.12.1

Establishing the Vibration Effect on Accuracy

A gun barrel is being designed to fire at a rate of 300 rounds per minute. It is comprised of three sections of length as given in Fig 1.12.1 ahead. Other necessary data is as follows:

Radius of sections:
A-B: 0.025 m
B-C: 0.022 m
C-D: 0.019 m
Mass density of steel: 7860.00 kg/m^3
Young's Modulus of steel: 1.8.10^{11} N/m^2
Bore radius: 0.00635 m

Investigate its vibration characteristics and determine whether the condition for accuracy of the weapon is achieved.

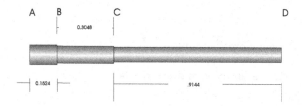

Fig 1.12.1: Dimensions of a given gun barrel

Solution

The investigation is carried out by the construction of a workbook in the following sequence:

100

Determination of Circular Speed under Static Loading

Areas and moments of inertia of the cross sections are calculated with the help of standard formulae.

Weight of each of the three sections is calculated from the given dimensions and mass density of the steel.

Bending moments are calculated at points A, B, C and D of the barrel.
Values of bending moment/moment of inertia are calculated at A, B and C.
Bending moments/moment of inertia at the mid sections AB, BC and CD are calculated.
The moment of the area of the bending moment diagram is calculated from left to right at each midsection.
Deflection at each midsection is calculated by dividing the moment of the moment of the area under the bending moment diagram by EI, according to the methods detailed in Article 1.11.
Circular speed is calculated with the help of Equation [1.11.6]

Spread sheet to determine circular speed: Static Loading					
Data					
Density of steel kg/m^2		7860	g m/s^2	9.815	
Inner radius m	0.00635	E Pa	1.8E+11		
Length of section m			Radius of section m		
A-B	B-C	C-D	A-B	B-C	C-D
0.1524	0.3048	0.9144	0.025	0.022	0.019
Area of section m^2			Moment of inertia of section m^4		
1.900E-03	1.425E-03	1.013E-03	3.26E-07	1.90E-07	1.02E-07
Weight of sections N					
22.34	33.51	71.49			
Bending moment					
A	B	C	D		
77.28	59.58	32.68	0.00		
Bending moment/I					
2.37E+08	3.13E+08	3.20E+08	0.00		
Bending moment /I at midsection					
2.75E+08	3.16E+08	1.60E+08	0.00		
Moment of area under bm diagram taken from left to right at mid section					
7.26E+05	2.25E+07	1.33E+08			
Y m					
4.03E-06	1.25E-04	7.38E-04			
WY			ΣWY		
9.01E-05	4.19E-03	5.27E-02	5.70E-02		
WY^2			sigmaWY^2		
3.63E-10	5.25E-07	3.89E-05	3.94E-05		
ω rad/s	119.1483				

Fig 1.12.2: Worksheet for determination of circular speed under static loading

Determination of Circular Speed under Dynamic Loading

The dynamic loads for each section are calculated using the first value of circular speed found from the equation for dynamic loading based on the static mass and acceleration vide Equation [1.11.4]. Making use of these dynamic loads, the first iteration for determination of the circular speed is carried out in an identical manner as in the case of static loading.

Spread sheet to determine circular speed: 1st Iteration dynamic loading					
Data					
Density of steel		7860	g	9.815	
Inner radius	0.00635	E	1.8E+11		
Length of section			Radius of section		
A-B	B-C	C-D	A-B	B-C	C-D
0.1524	0.3048	0.9144	0.025	0.022	0.019
Area of section			Moment of inertia of section		
1.900E-03	1.425E-03	1.013E-03	3.26E-07	1.90E-07	1.02E-07
Weight of sections					
0.13	6.07	76.26			
Bending moment					
A	B	C	D		
71.59	59.04	34.87	0.00		
Bending moment/I					
2.20E+08	3.10E+08	3.41E+08	0.00		
Bending moment /I at midsection					
2.65E+08	3.26E+08	1.71E+08			
Moment of area under bm diagram					
6.82E+05	2.18E+07	1.36E+08			
Y					
3.79E-06	1.21E-04	7.53E-04			
WY			ΣWY		
4.94E-07	7.34E-04	5.74E-02	5.82E-02		
WY^2			ΣWY^2		
1.87E-12	8.88E-08	4.33E-05	4.34E-05		
ω1	114.7606				

Fig 1.12.3: Worksheet for 1st iteration for calculation of circular speed under dynamic loading

The iteration is continued making use of the value of circular speed obtained from the previous worksheet until acceptably close values of circular speed are obtained from two successive iterations.

Spread sheet to determine circular speed: 2nd Iteration dynamic loading					
Data					
Density of steel		7860	g	9.815	
Inner radius	0.00635	E	1.8E+11		
Length of section			Radius of section		
A-B	B-C	C-D	A-B	B-C	C-D
0.1524	0.3048	0.9144	0.025	0.022	0.019
Area of section			Moment of inertia of section		
1.900E-03	1.425E-03	1.013E-03	3.26E-07	1.90E-07	1.02E-07
Weight of sections					
0.11	5.44	72.25			
Bending moment					
A	B	C	D		
67.73	55.88	33.03	0.00		
Bending moment/I					
2.08E+08	2.94E+08	3.23E+08	0.00		
Bending moment /I at midsection					
2.51E+08	3.08E+08	1.62E+08			
Moment of area under bm diagram					
6.45E+05	2.06E+07	1.28E+08			
Y					
3.58E-06	1.15E-04	7.13E-04			
WY			ΣWY		
4.07E-07	6.23E-04	5.15E-02	5.22E-02		
WY^2			ΣWY^2		
1.46E-12	7.13E-08	3.68E-05	3.68E-05		
ω2	117.8976				

Fig 1.12.4: Worksheet for 2nd iteration to determine circular speed under dynamic loading.

Spread sheet to determine circular speed: 3rd Iteration dynamic loading					
Data					
Density of steel kg/m^2		7860	g	9.815	
Inner radius	0.00635	E	1.8E+11		
Length of section			Radius of section		
A-B	B-C	C-D	A-B	B-C	C-D
0.1524	0.3048	0.9144	0.025	0.022	0.019
Area of section			Moment of inertia of section		
1.900E-03	1.425E-03	1.013E-03	3.26E-07	1.90E-07	1.02E-07
Weight of sections					
0.11	5.43	72.21			
Bending moment					
A	B	C	D		
67.69	55.85	33.01	0.00		
Bending moment/I					
2.08E+08	2.93E+08	3.23E+08	0.00		
Bending moment /I at midsection					
2.51E+08	3.08E+08	1.62E+08			
Moment of area under bm diagram					
6.45E+05	2.06E+07	1.28E+08			
Y					
3.58E-06	1.14E-04	7.13E-04			
WY			ΣWY		
4.06E-07	6.22E-04	5.15E-02	5.21E-02		
WY^2			ΣWY^2		
1.46E-12	7.12E-08	3.67E-05	3.68E-05		
ω3	117.9311				

Fig 1.12.5: Worksheet for 3rd iteration for determination of circular speed under dynamic conditions

Spread sheet to determine circular speed: 4th Iteration dynamic loading					
Data					
Density of steel kg/m^2		7860	g	9.815	
Inner radius	0.00635	E	1.8E+11		
Length of section			Radius of section		
A-B	B-C	C-D	A-B	B-C	C-D
0.1524	0.3048	0.9144	0.025	0.022	0.019
Area of section			Moment of inertia of section		
1.900E-03	1.425E-03	1.013E-03	3.26E-07	1.90E-07	1.02E-07
Weight of sections					
0.11	5.43	72.21			
Bending moment					
A	B	C	D		
67.69	55.85	33.01	0.00		
Bending moment/I					
2.08E+08	2.93E+08	3.23E+08	0.00		
Bending moment /I at midsection					
2.51E+08	3.08E+08	1.62E+08			
Moment of area under bm diagram					
6.45E+05	2.06E+07	1.28E+08			
Y					
3.58E-06	1.14E-04	7.13E-04			
WY			ΣWY		
4.06E-07	6.22E-04	5.15E-02	5.21E-02		
WY^2			ΣWY^2		
1.46E-12	7.12E-08	3.67E-05	3.68E-05		
ω4	117.9315				
fn	18.7694	fr		5	
Fr	3.7539				

Fig 1.12.6: Worksheet for 4th iteration for determination of circular speed under dynamic conditions, natural frequency and Frequency Ratio

An acceptable value of circular speed has been established by the fourth iteration. The change in circular speed from the 3rd iteration to the 4th iteration being less than .01%.

Using this accepted value of the circular speed, the natural frequency of the barrel is obtained from Equation [1.11.7].

Finally the Frequency Ratio is calculated with the help of Equation [1.11.1]. This works out to 3.75.

The value of the Frequency Ratio is seen against the curve on the accuracy-Frequency Ratio curve. The criterion earlier stated is met as a Frequency Ratio of 3.75 is achieved is entirely acceptable from the accuracy point of view.

2

Design of Breech Assemblies

2.1 Design Considerations of Breech Assemblies

Functions and Components of a Breech Assembly

The breech assembly consists of the components responsible for closing and sealing of the breech end of the gun barrel, after the ammunition has been loaded, the firing mechanism which initiates the explosive train of the propellant system, the mechanism for opening of the breech after firing and the means for removal of the unexpended remnants of ammunition, if any. It also incorporates necessary arrangements for ensuring safety during the firing sequence. In order to fulfill the aforementioned functions, a typical breech assembly consists of the breech ring which attaches the breech assembly to the barrel and carries a breechblock or screw, an operating mechanism which imparts opening and closing motion to the block or screw, an electric or percussion firing mechanism, means of obturation, extracting and ejecting mechanisms, necessary safety devices and stops to limits movements.

Sequence of Operation of a Breech Assembly in Outline

Starting from the breech open condition, the sequence of events involving the breech assembly is as follows. The ammunition is fed into the breech end of the barrel manually or by automated loading devices. The breech aperture is then closed by lateral movement of the breechblock or by rotary motion of the breech screw, within the breech ring. In the final stage of motion, locking of the breechblock or screw takes place. On firing, the explosive train of the propellant system is initiated either by percussion or by passage of electric current through the primer. High pressure gases produced as a result of the combustion of the propellant are prevented from rearward leakage. This is realized, as the case may be, by the lips of the cartridge case expanding to eliminate the cartridge case-chamber clearance, or by the sealing effect of the obturator pad against the chamber walls. Leakage through the initiating device is eliminated by obturating means incorporated in the initiating device itself. In accordance with the conservation of momentum law, kinetic energy is imparted to the barrel and its attachments. Some of this energy of recoil is stored in the recuperator which serves to return the gun to the firing position. During counter recoil it is possible to utilize some of the stored energy to activate the operating mechanism for unlocking, cocking of the firing mechanism, opening of the breech, extraction and ejection of the spent case and storing of sufficient energy in a spring for closure of the breech upon reloading. In the case of manually operating mechanisms, the opening of the breech, extraction and ejection is done after the gun returns to the firing position.

Selection of a Breech Assembly

The choice of a breech assembly for large calibre weapons, out of the two available options, namely the sliding block and the breech screw depends on important factors as detailed ahead.

Weapon configuration implies the influence of space and weight constraints and availability of power for actuating functions of the breech assembly's mechanisms. In self propelled artillery and tank turrets, while auxiliary power is available, space is crucial. Here a greater degree of automation to assist the crew is desirable. With towed artillery space is not a limiting aspect on the efficient functioning of the crew.

Ammunition is a major deciding factor with respect to the choice of a breech assembly. Ammunition for large calibre weapons may be based on a metal cartridge case which holds the propellant charge and primer. The case may be crimped to the projectile or separate from it. From the breech assembly point of view no material difference arises from either arrangement. Alternately, the propellant may be contained in bags of combustible material. In this instance, the primer is separately inserted into the firing mechanism after the breech is closed and locked. For ammunition systems with metal cartridge cases, the sliding block option is the obvious choice. For bag loaded charges, the breech screw arrangement, with an integral obturating device is preferred. Typical layouts of both systems are depicted in the figure below.

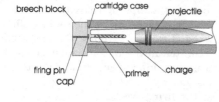

Fig 2.1.1: Metal cartridge case and bagged charge ammunition systems shown loaded.

Sliding Block Mechanisms

The breech ring is the outer structural component. It has an axial boring to facilitate loading and a lateral boring which provides surfaces over which the breechblock slides, the limits of its movement corresponding to the open and closed positions of the breech. Movement of the breechblock is achieved through the operating mechanism which may be manual or semi automatic in nature. The sliding block motion may be in either in the vertical or horizontal plane. In both cases the movement of the block is linear in a single cycle. This is the main advantage of the sliding block with respect to semi automatic operation. The block travel is usually reduced by suitably cutting out the end of the sliding block so that the total travel distance of the block is only a little more than the chamber diameter.

The sliding block also employs the wedge principle to ensure proper seating of the cartridge case before firing. This is achieved by sloping both the grooves in the breech ring and the corresponding contact surfaces of the block by a small angle, usually in the region of a degree. As the block slides to the closed position, it also moves forward marginally in the axial direction to fully support the cartridge case base. The front face of the breechblock is beveled to thrust the cartridge case into the chamber as it closes.

To prevent jamming on firing, a clearance, known as cartridge head space, is catered for between the front face of the breechblock and the base of the cartridge case. These aspects are illustrated in the figure ahead.

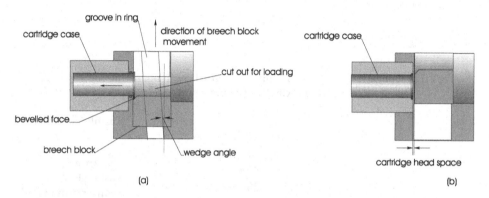

Fig 2.1.2: (a) Breechblock in the process of closing showing action of beveled face. (b) Breechblock in closed position showing cartridge headspace (exaggerated).

Semi Automatic Operating Mechanisms

Cam Operation

The operation of a typical semi automatic operating mechanism is depicted in Fig 2.1.3. During recoil, the spring loaded operating cam is thrust aside by the beveled face of the cam follower splined to the crankshaft of the breech operating mechanism. During counter recoil, however, the follower contacts the cam working surface head on and is forced to climb over it, imparting rotation to the crankshaft of the operating mechanism. Further detailed explanation and analysis is covered in Article 2.5 of this chapter.

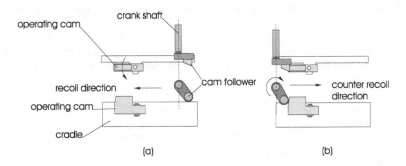

Fig 2.1.3: Operation of a typical semi automatic breech operating mechanism.
(a) During recoil. (b) During counter recoil

Crank Operation

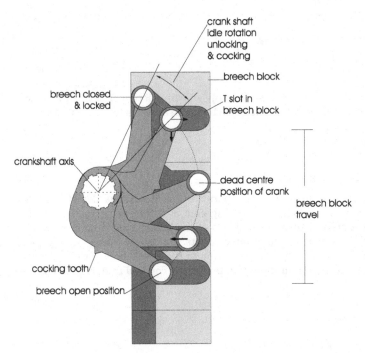

Fig 2.1.4: Crankshaft & crank rotation and breechblock motion

Crankshaft and crank rotation, and breechblock motion is depicted in Fig 2.1.10. The first angular motion of the crank serves to move the crank roller within the sloped portion of the T slot in the breechblock. This unlocks the

112

breechblock and cocks the firing mechanism by the thrust of the crank cocking tooth acting on the firing mechanism cocking lever. The vertical movement of the breechblock begins when the crank roller contacts the lower horizontal surface of the T slot in the breechblock. The crank roller continues its downwards descent and movement to the right until it reaches the dead centre position. After the dead centre position it continues its descent but changes sense in that it now moves to the left. As far as the breechblock is concerned, the movement is downward throughout. At the same time the rotation of the crankshaft turns the closing spring lever splined to it, thereby compressing the closing spring. The crank action ceases when the cam follower reaches the horizontal surface of the operating cam. Closing of the breech is achieved by rotation of the crankshaft in the opposite direction under the influence of the closing spring, resulting in rotation of the crank and thrust of the crank roller on the upper horizontal surface of the T slot.

Breechblock Locking

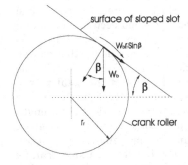

Fig 2.1.5: Breech locking action of crank roller and sloped portion of T slot in breechblock

W_b: weight of breechblock
r_r: crank roller radius
β : slot angle with horizontal

Breechblock locking is achieved by the sloped slot contiguous to the T slot of the breechblock. The moment the crank roller enters the sloped slot during closing, the weight of the breechblock induces a moment about the centre of the roller which tends to drive it up the slope and further into the sloped slot and the crank into the breech closed position. This is illustrated in Fig 2.1.11.

Breech Screw Mechanisms

In the breech assemblies with breech screws, the breech closing is based on the principle of the screw thread. The breech ring is threaded internally to receive the breech screw, which has matching teeth on its outer surface. To achieve speed of operation and reduce the angular rotation necessary for full mating of the teeth on the screw with the threads in the ring, the threading is alternated having segments of threads followed by equal plane segments. Mating can thus be achieved by rotation of the screw through an angle corresponding to the arc of a segment. The thread pitch is selected so that friction force prevents the screw from rotating during the period of activity of the propellant gas pressure. As a further measure of safety, additional locking devices are usually incorporated. Important components of a typical breech screw assembly are described in the paragraphs ahead.

Types of Interrupted Thread Breech Screws

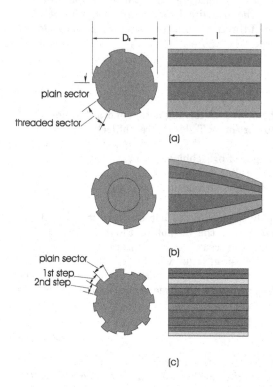

Fig 2.1.6: (a) Simple threaded breech screw. (b) Tapered conical breech screw. (c) Welin breech screw.

The earliest breech screw had simple threaded sectors alternating between plane sectors. With this kind of arrangement, the mating surface area of the threads was limited to 50% of the total thread surface area, the length of the screw being fixed by strength considerations related to a given maximum propellant gas force. However, the rotation angle can be decreased by increasing the number of sectors but within practical limits. If the number of sectors was increased, the length of the screw increased proportionately. This increased length called for three distinct motions during opening and closing, i.e. the motions of rotation, axial withdrawal and swinging clear to one side to enable access to the chamber.

This drawback of increase in breech screw length is somewhat lessened in the conically tapered breech screw. Here, the thread engagement area is increased by significantly increasing the breech screw diameter towards the rear of the screw, resulting in more favourable thrust distribution. This arrangement however increased the complexity of manufacture of the screw.

Some current breech screw designs employ the Welin Screw. In this design, the threaded sector of the screw is broken down sequentially into two or more threaded sub-sectors of increasing diameter. The angle subtended by a threaded sub sector and the plain sector being equal. This design affords greater engagement surface than the screw with simple threaded sector. For a fixed load, the length of the screw can be reduced. A shorter screw is advantageous in that the screw can be swung clear of the breech ring without axial withdrawal. This considerably simplifies the operating mechanism design and implementation. The Welin breech screw with matching breech ring, though technically lucrative, poses complexities of manufacturing which may in many cases outweigh its technical returns.

Breech Screw Operating Mechanisms

The operating mechanism of a breech screw assembly differs from that of the sliding block mechanism in that at least two distinct motions occur before the breech is accessible for loading. In the older models, the three movements taking place were the anti clockwise rotation of the screw, withdrawal in the direction of the bore axis and swinging clear to one side. Current practice, made possible by shorter breech screws, consists of rotary motion followed immediately by swinging clear to one side.

The rotary motion of the breech screw is effected by the interaction of a mechanism consisting of the breech lever, the crankshaft and the cross head which moves horizontally within a cylindrical recess in the breech screw. When the breech-operating lever is swung backwards in the vertical plane, the crankshaft rotates about its axis. The cross head which is fitted to the crank moves within the cylindrical recess of the breech screw, turning the screw in the anti clockwise direction. The movements may be understood by correlation with the breech screw operating mechanism depicted exploded in Fig 2.1.6.

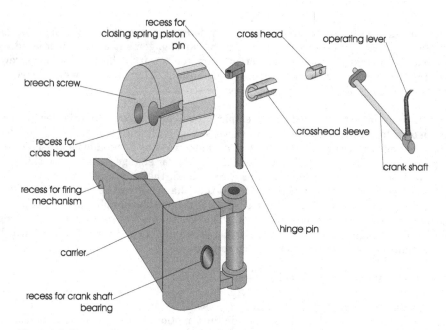

Fig 2.1.7: Essential components of breech screw operating mechanism shown exploded

The second motion in the breech screw opening cycle takes place when the operating lever is pulled to the right. Now, the carrier, which is pivoted about the hinge axis, rotates about it swinging the breech screw to the open position to the right. The two distinct motions i.e. rotation of the block and swinging clear are smoothed into one motion by the action. The roller on the breech screw contacts the curved surface of the cam on the breech ring face, which turns the direction of the roller from that corresponding to breech screw rotation, through a right angle into the direction of swing free about the hinge pin axis. The cam action ensures that the breech opening and closing is done smoothly and without impact to the equipment and discomfort to the operator. The cam action is depicted in Fig 2.1.8.

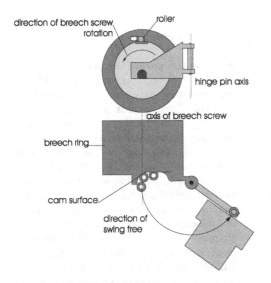

direction of breech screw rotation

roller

hinge pin axis

axis of breech screw

breech ring

cam surface

direction of swing free

Fig 2.1.8: Transition of breech screw rotation to swing free motion by cam action.

Closing Spring Assembly

A closing spring assembly is provided which aids the human effort during closing of the breech. It is also designed to hold the breech screw in the open position for loading and inspection. A typical closing spring assembly consists of a spring in compression between a piston and one end inner surface of its cylindrical container. The piston rod is pinned to the hinge pin crank and the far end of the cylinder is pivot fitted onto the breech ring. As the breech screw swings free, the hinge pin crank gets rotated pulling the piston rod against compression of the spring. The maximum compression of the spring is made to occur slightly before the breech fully open position is reached. After this the spring expands a little thereby holding the breech screw in the open position. The action of the closing spring is illustrated in Fig 2.1.9 ahead.

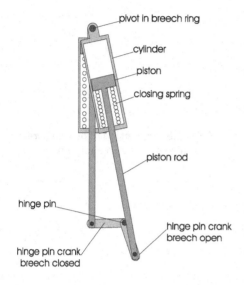

Labels on figure:
pivot in breech ring
cylinder
piston
closing spring
piston rod
hinge pin
hinge pin crank
breech open
hinge pin crank
breech closed

Fig 2.1.9: Action of the closing spring assembly

Firing Mechanisms

An essential part of the breech assembly is the firing mechanism. Firing mechanisms may be mechanical or electric. For obvious reasons, electric firing mechanisms invariably incorporate a mechanical back up. The nature of explosives in the primer and the type of firing mechanism are closely interrelated. Explosives used with electric firing mechanisms are initiated by the heat resulting from the passage of electric current through them. Primers initiated mechanically by the heat caused by friction or impact are called percussion primers. Firing mechanisms are commonly classified as continuous pull, hammer, inertia percussion and continuous pull in-line hammer types. A brief description of each type follows.

Continuous pull firing mechanisms are common employed with howitzers and field artillery pieces. Here the cocking and release of the firing mechanism takes place as a result of one continuous pull on a lanyard or other activating device. The action of the lanyard firstly compresses the firing spring and retains the firing pin in the cocked position. Further pull of the lanyard trips the sear allowing the spring to expand and drive the firing pin to impact with the primer.

The hammer type of firing mechanism, as the name suggests is based on a spring loaded hammer, which rotates about a hinge. The hammer is cocked and held in the cocked position. The firing pin is always held in the retracted position by spring force. Upon trigger action, the hammer rotates to strike the base of the firing pin which impacts it against the primer against the force of the firing pin. After impact, the firing pin is returned to the retracted position by the firing pin spring.

Continuous pull in line hammer firing mechanisms follow the same principle as the hammer type, in this case however, the hammer is not hinged but moves in the axial direction rearward under the pull from the activating device, compressing the main spring where it is cocked. Further pull trips the sear which releases the hammer to move forward and strike the firing pin base.

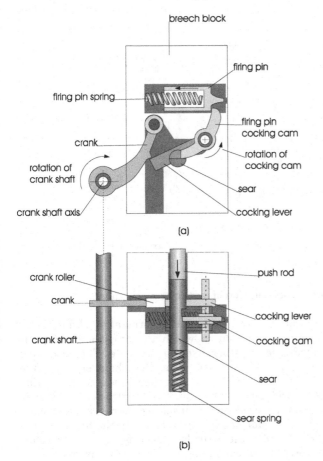

(a)

(b)

Fig 2.1.10: Action of spring percussion type firing mechanism. (a) During cocking. (b) During firing.

A spring actuated percussion type of firing mechanisms is compatible with semi automatic breech operating mechanisms. A description of a typical such mechanism follows. During counter recoil, cam action rotates the crankshaft, which in turn rotates the crank. The cocking tooth of the crank thrusts against the cocking lever, which in turn rotates the firing pin cocking cam. The firing pin cocking cam compresses the firing spring and at the extremity of its rotation gets held by the sear, restraining the firing pin in the cocked position against the pressure of the compresses firing spring.

Trigger action causes the sear to move in a direction

119

along its longer axis against the force of the sear spring. The cut out part of the sear comes in front of the firing pin cocking cam, tripping the cocking cam. The firing spring expands to impact the firing pin against the primer. The sequence of operation is depicted in Fig 2.1.10.

Extracting Mechanisms

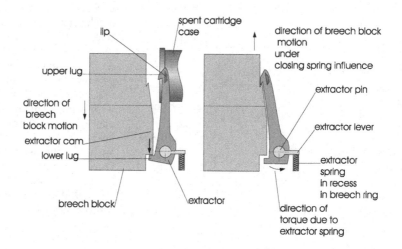

Fig 2.1.11: Sequence of operation of extractor mechanism

Extracting mechanisms are associated with equipments incorporating sliding block breech assemblies and employing ammunition in which the propellant charge is contained in a metallic cartridge case, whether fixed or separate. The extracting mechanism functions to extract the spent case from the chamber, hold the breechblock in the open position and release the compression of the closing spring when tripped by action of the cartridge case rim against the extractor lips during loading. Normally a pair of extractors is used, although space constraints may dictate the use of a single extractor. The extractor lips fit into recesses in the breech end of the chamber face. These lips lie forward of the cartridge case rim when the breech is closed. The extractors also have projections, called lugs, facing rearwards, both at the upper and lower extremities. The upper lugs serve to engage with projections on the breechblock extractor cam in order to retain the breechblock in the breech open position. The lower lugs are struck by the lower end of the same extractor cam at the end of the downward movement of the breechblock thereby energetically rotating the extractors about the extractor pin axis, to extract and eject the spent cartridge case in one motion. The extractor pin is spring loaded with the help of a spring and lever arrangement, which induces the pin and the extractors splined to it to rotate in the counterclockwise direction unless otherwise constrained.

As the breechblock falls, the lower lugs of the extractor ride on the surface of the extractor cam imparting counterclockwise motion about the extractor pin axis to the extractors. The extractor lips now propel the cartridge case out of the open breech end of the chamber. At the fully open position, cam action ceases but the extractor pin spring force causes the extractors to snap over the projections at the top of the extractor cams, maintaining the breech in the open position. On loading, the cartridge case rim strikes the extractor lips displacing them from the cam projections and releasing the breechblock, which rises to close the breech under the influence of the closing spring. The sequence of operation is depicted in Fig 2.1.11.

2.2 Stresses in Interrupted Thread Breech Screws

Stress Analysis in Breech Screws

Light weight of components of a weapon system is always a paramount design consideration. In the case of a breech screw, the diameter is fixed, more or less by the calibre of the weapon. Based on stress analysis it is nevertheless possible to limit the axial length of the screw and thus minimize its weight. Because of the complicated geometry, loading and boundary conditions, an interrupted breech screw presents an intricate stress analysis problem, which cannot be generalized. However a basic approach based on the theory of thick circular plates, with some simplifying assumptions, does yield constructive information.

Consider a breech screw to be a thick circular plate of thickness h, fixed along it periphery and loaded symmetrically with respect to its axis. Deformation, displacements and stresses will be in this case symmetrical with respect to the axis of the plate. With reference to Fig 2.2.1, an element of radial thickness dr at a radial distance r from the axis of the plate is isolated. Also the sides of the element subtend an angle $d\phi$ at the centre.

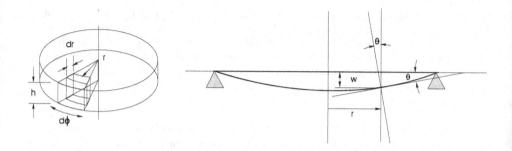

Fig 2.2.1: Bending of a symmetrically loaded thick circular plate

Taking the origin of the reference system as the intersection of the middle surface of the plate and its axis, the deflection of the plate is w downwards and the angle of rotation of the normal to the middle surface is θ, where w and θ are functions only of the radius r. For small deflections, θ, w and r are related by the expression:

$$\theta = -\frac{dw}{dr} \qquad\qquad\qquad\qquad\qquad\qquad\qquad\qquad\qquad [2.2.1]$$

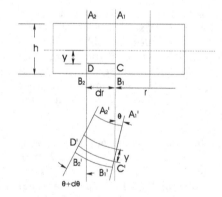

Fig 2.2.2: Axial section of thick circular plate

With reference to Fig 2.2.2, depicting an axial section of the plate, the normals A_1, B_1 and A_2, B_2 before bending rotate to A_1', B_1' and A_2', B_2' after bending by angles θ and $\theta + d\theta$ respectively. The radial elongation of the segment CD located at a distance y from the middle surface is:

$$y(\theta + d\theta) - y\theta = yd\theta$$

122

The radial strain is given by:

$$\varepsilon_r = y\frac{d\theta}{dr} \qquad\qquad\qquad\qquad [2.2.2]$$

Before bending, the length of the circumference through C was $2\pi r$ and after bending it is $2\pi(r + y\theta)$. Hence the circumferential strain is:

$$\varepsilon_t = y\frac{\theta}{r} \qquad\qquad\qquad\qquad [2.2.3]$$

Expressing the stresses in terms of the strains, the radial stress is:

$$\sigma_r = \frac{E}{1-\mu^2}\left(\varepsilon_r + \mu\varepsilon_t\right) \qquad\qquad\qquad [2.2.4]$$

The hoop stress is:

$$\sigma_t = \frac{E}{1-\mu^2}\left(\varepsilon_t + \mu\varepsilon_r\right) \qquad\qquad\qquad [2..2.5]$$

Making use of Equations [2.2.2] and [2.2.3]:

$$\sigma_r = \frac{Ey}{1-\mu^2}\left(\frac{d\theta}{dr} + \mu\frac{\theta}{r}\right) \qquad\qquad\qquad [2.2.6]$$

$$\sigma_t = \frac{Ey}{1-\mu^2}\left(\frac{\theta}{r} + \mu\frac{d\theta}{dr}\right) \qquad\qquad\qquad [2.2.7]$$

Apart from normal stresses, the faces of the element may be acted upon by shearing stresses. Because of symmetry, these will act on planes perpendicular to the radius and in the vertical direction only. Considering the equilibrium of the element in isolation, let the shear force per unit circumferential length be denoted by Q. Hence the shearing force on the inner face of the element is $Qrd\phi$ and on the outer face it will be $(Q+dQ)(r+dr)d\phi$. Since the stress in the top and bottom layers are identical with sign reversed, normal forces on the faces of the element are absent. The normal stresses on the faces reduce to resultant moments in vertical planes. Fig 2.2.3 refers.

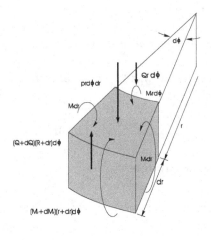

Fig 2.2.3: Moments and shear forces on element of circular plate

The resultant moment on the on the inner face is:

$$M_r r d\phi = r d\phi \int_{-\frac{h}{2}}^{\frac{h}{2}} \sigma_r y \, dy$$

Making use of Equation [2.2.6]:

$$M_r = \frac{E}{1-\mu^2}\left(\frac{d\theta}{dr} + \mu\frac{\theta}{r}\right)\int_{-\frac{h}{2}}^{\frac{h}{2}} y^2 dy = \frac{Eh^3}{12(1-\mu^2)}\left(\frac{d\theta}{dr} + \mu\frac{\theta}{r}\right)$$

Or:

$$M_r = D\left(\frac{d\theta}{dr} + \mu\frac{\theta}{r}\right) \dots\dots\dots\dots\dots\dots\dots\dots\dots\dots\dots\dots\dots\dots\dots\dots\dots\dots\dots\dots[2.2.8]$$

Where D is called the flexural rigidity of the plate, given by:

$$D = \frac{Eh^3}{12(1-\mu^2)} \dots\dots\dots\dots\dots\dots\dots\dots\dots\dots\dots\dots\dots\dots\dots\dots\dots\dots\dots\dots\dots\dots\dots[2.2.9]$$

Similarly, making use of Equation [2.2.7], the moment acting on the side face of the element:

$$M_t dr = dr \int_{-\frac{h}{2}}^{\frac{h}{2}} \sigma_t y \, dy = \frac{E}{1-\mu^2}\left(\frac{\theta}{r} + \mu\frac{d\theta}{dr}\right)\int_{-\frac{h}{2}}^{\frac{h}{2}} y^2 dy = \frac{Eh^3}{12(1-\mu^2)}\left(\frac{\theta}{r} + \mu\frac{d\theta}{dr}\right)$$

Finally:

$$M_t = D\left(\frac{\theta}{r} + \mu\frac{d\theta}{dr}\right)$$ ....[2.2.10]

The external forces applied to the element also include the propellant gas force $p_m rd\phi dr$. Here p_m is the maximum propellant gas pressure. Summing up the force in the direction of the axis of the plate:

$$(Q + dQ)(r + dr)d\phi - Qrd\phi - p_m rd\phi dr = 0$$

From which:

$$p_m r = \frac{d(Qr)}{dr}$$ ....[2.2.11]

$\because \dfrac{d(Qr)}{dr} = Q + r\dfrac{dQ}{dr}$ and neglecting smaller quantities.

Summing up the moments of all forces with respect to an axis tangential to the inner face in the middle plane:

$$(M_r + dM_r)(r + dr)d\phi - M_r rd\phi - p_m rdrd\phi\frac{dr}{2} - 2M_t dr\frac{d\phi}{2} + (Q + dQ)(r + dr)d\phi dr = 0$$

Neglecting smaller quantities of higher order:

$$M_r + \frac{dM_r}{dr}r - M_t + Qr = 0$$

Substituting for M_r and M_t from Equations [2.2.8] and [2.2.9] in the above expression:

$$\frac{d^2\theta}{dr^2} + \frac{1}{r}\frac{d\theta}{dr} - \frac{\theta}{r^2} = -\frac{Q}{D}$$

For ease of integration, this may be written in the form:

125

$$\frac{d}{dr}\left[\frac{1}{r}\frac{d}{dr}(\theta r)\right] = -\frac{Q}{D} \quad\text{...[2.2.12]}$$

Writing θ in terms of w and r, this can also be written as:

$$\frac{d}{dr}\left[\frac{1}{r}\frac{d}{dr}\left(r\frac{dw}{dr}\right)\right] = \frac{Q}{D} \quad\text{...[2.2.13]}$$

When the breech screw of radius a is subject to the maximum propellant gas pressure p_m over its front face, then the shearing force Q per unit distance at a radial distance r is given by:

$$2\pi rQ = \pi r^2 p_m$$

Or:

$$Q = \frac{p_m r}{2} \quad\text{...[2.2.14]}$$

Substituting in Equation [2.2.12] and integrating:

$$\frac{1}{r}\frac{d}{dr}\left(r\frac{dw}{dr}\right) = \frac{p_m r^2}{4D} + C_1$$

Multiplying both sides by r:

$$\frac{d}{dr}\left(r\frac{dw}{dr}\right) = \frac{p_m r^3}{4D} + C_1 r$$

Integrating the expression immediately above:

$$r\frac{dw}{dr} = \frac{p_m r^4}{16D} + \frac{C_1}{2}r^2 + C_2$$

126

From which:

$$\frac{dw}{dr} = \frac{p_m r^3}{16D} + \frac{C_1}{2} r + \frac{C_2}{r} \quad\text{...}[2.2.15]$$

C_1 and C_2 are constants of integration to be determined from the conditions at the centre and edge of the breech screw.

Recalling that the breech screw is a thick circular plate with edges fixed, the slope of the deflection curve in the radial direction is zero at $r = a$ and $r = 0$. Hence from Equation [2.2.14]:

$$\frac{p_m r^3}{16D} + \frac{C_1}{2} r + \frac{C_2}{r} = 0 \text{, from which } C_2 = 0.$$

Also:

$$\frac{p_m a^3}{16D} + \frac{C_1}{2} a = 0 \text{ or } C_1 = -\frac{p_m a^2}{8D}$$

Substituting for C_1 and C_2 in Equation [2.2.15]:

$$\theta = -\frac{dw}{dr} = \frac{p_m r}{16D}\left(a^2 - r^2\right) \quad\text{...}[2.2.16]$$

Differentiating the expression immediately above:

$$\frac{d\theta}{dr} = \frac{p_m}{16D}\left(a^2 - 3r^2\right)\text{...}[2.2.17]$$

Comparing Equations [2.2.6], [2.2.7] and [2.2.8], [2.2.10], the stresses in terms of the bending moments are given by:

$$\sigma_r = \frac{12M_r}{h^3} y \text{ and } \sigma_t = \frac{12M_t}{h^3} y \quad\text{...}[2.2.18]$$

The maximum stresses occur at $y = \pm\frac{h}{2}$:

$$\sigma_{r(max)} = \frac{6M_r}{h^2} \text{ and } \sigma_{t(max)} = \frac{6M_t}{h^2} \dots\dots\dots\dots\dots\dots\dots\dots\dots\dots\dots\dots\dots\dots\dots\dots\dots\dots\dots[2.2.19]$$

In order to determine the stresses, M_r and M_t can be evaluated with the help of Equations [2.2.16] and [2.2.17], which define θ and $\frac{d\theta}{dr}$.

2.3 Estimation of Stresses in an Interrupted Breech Screw

Design Exercise 2.3.1

It is desired to compute the radial, hoop and equivalent stresses in an interrupted breech screw of radius 80 mm and thickness 150 mm given that the maximum propellant gas pressure acting uniformly over the front face of the screw is 227.0 MPa. Young's Modulus for the steel used is 207.0 G Pa. Poisson's ratio is 0.3. The breech screw may be assumed a thick plate clamped all along its periphery. The effect due to the central hole in the breech screw may also be assumed negligible. Compare the stress pattern for an identical screw of thickness reduced by 50 mm.

Solution

The stresses in the interrupted thread breech screw are estimated using the assumptions and analysis detailed in the foregoing article. The procedure adopted is as follows:

The flexural rigidity D is calculated with the help of Equation [2.2.9]. Next the radial and tangential bending moments are calculated from Equations [2.2.8] and [2.2.10], inserting the values of θ and $\frac{d\theta}{dr}$ from Equations [2.2.16] and [2.2.17]. Finally the maximum radial stress and hoop stress are computed from Equation [2.2.19]. A Matlab programme is written to compute and plot the results. The computer programme is given below followed by the graphical output.

Matlab Programme 2.3.1

% programme to compute stresses: interrupted thread breech screw
% file name breech screw.txt

```
a=.08 % breech screw diameter m
h1=.15 % breech screw thickness m
h2=.1 % reduced thickness of screw
E=207*10.^9 % youngs modulus Pa
mu=0.3 % Poissons ratio
pm=227.*10.^6 % max propellant gas pressure Pa
r=linspace(0,a,1000)

% thickness 0.15

D1=E.*h1.^3./(12.*(1-mu.^2)) % flex rigidity
sigyp=280.*10.^6 % yield strength Pa

theta1=pm.*r./(16.*D1).*(a.^2-r.^2)
dtheta1=pm./(16.*D1).*(a.^2-3.*r.^2)

Mr1=D1.*(dtheta1+mu.*theta1./r)
Mt1=D1.*(theta1./r+mu.*dtheta1)

sigr1=12.*Mr1./h1.^2
sigt1=12.*Mt1./h1.^2

sigeq1=sqrt(sigr1.^2-sigr1.*sigt1+sigt1.^2)

% thickness 0.1

D2=E.*h2.^3./(12.*(1-mu.^2)) % flex rigidity
theta2=pm.*r./(16.*D2).*(a.^2-r.^2)
dtheta2=pm./(16.*D2).*(a.^2-3.*r.^2)

Mr2=D2.*(dtheta2+mu.*theta2./r)
Mt2=D2.*(theta2./r+mu.*dtheta2)

sigr2=12.*Mr2./h2.^2
sigt2=12.*Mt2./h2.^2

sigeq2=sqrt(sigr2.^2-sigr2.*sigt2+sigt2.^2)
```

subplot('position',[.35,.55,.3,.3]),plot(r,sigr1,r,sigt1,r,sigeq1, r,sigyp)
title(' Stresses breech screw')
xlabel('radius m');ylabel('stress Pa')
gtext('radial stress');gtext('hoop stress');gtext('equivalent stress')
gtext('yield stress')

subplot('position',[.35,.1,.3,.3]),plot(r,sigr2,r,sigt2,r,sigeq2,r,sigyp)
title(' Screw of reduced thickness')
xlabel('radius m');ylabel('stress Pa')
gtext('radial stress');gtext('hoop stress');gtext('equivalent stress');gtext('yield stress')

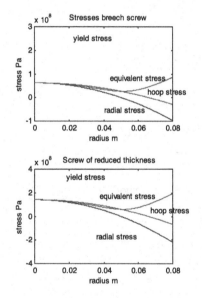

Fig 2.3.1: Stresses in interrupted thread breech screw

2.4 Stresses in Screw Threads of Breech Screws

Stresses on the threads of an interrupted breech screw are related to the design of the breech screw. The breech screw may be segmented or consist of stepped segments and plane segments. In all cases the net thrust on the existing threads can not be less than the propellant gas force on the breech screw face. The propellant gas force on the breech screw face is given by:

$$P = p_m \pi a^2 \quad \dots \dots \dots \dots \dots \dots \dots \dots \dots \dots \dots \dots \dots \dots \dots \dots \dots \dots \dots \dots \dots \dots \dots \dots \dots \dots [2.4.1]$$

p_m: maximum propellant gas pressure
a: radius of breech screw face

Assuming that the outer circumference of the breech screw is equally shared by the threaded and the interrupted segments, the shear area per thread, is given by:

$$A = \frac{\pi}{3} (d_1 + d_2)(p - \lambda)$$

d_1: diameter of larger step
d_2: diameter of smaller step
p: thread pitch
λ: tooth clearance

If n threads or in contact, then the net shear area is:

$$A = \frac{\pi n}{3} (d_1 + d_2)(p - \lambda) \quad \dots \dots \dots \dots \dots \dots \dots \dots \dots \dots \dots \dots \dots \dots \dots \dots \dots \dots \dots [2.4.2]$$

It follows that the mean shear stress on the threads is $\tau_m = \dfrac{P}{A}$

Or:

$$\tau_m = \frac{3 p_m a^2}{n(d_1 - d_2)(p - \lambda)} \quad \dots \dots \dots \dots \dots \dots \dots \dots \dots \dots \dots \dots \dots \dots \dots [2.4.2]$$

2.5 Semi Automatic Breech Mechanisms

Sequence of Operation of Semi Automatic Breech Mechanisms

On firing, the thrust due to propellant gas pressure acts on the base of the cartridge case is conveyed to the breech ring via the breechblock. The breech ring is attached to the barrel by means of the barrel nut. As a result, the entire assembly consisting of the barrel the breech assembly and other attachments displace backwards. The operating crankshaft has the arm of the breech-operating follower splined to it. During recoil displacement, the beveled edge of the follower encounters the spring-loaded operating cam, which is fixed to the cradle. The cam is deflected outwards and the cam follower continues its rearward displacement along with the rest of the recoiling parts. During counter recoil, the follower contacts the operating cam surface head on. As the recoiling parts are in motion, the follower rides over the cam. The interaction of the follower and cam rotates the crankshaft clockwise about its axis. Also splined to the crankshaft is the breechblock crank, the circular roller of which rides in the T slot of the breechblock.

The breech-operating crank rotates about the axis of the crankshaft. As the roller moves along the inclined portion of the T slot in the breechblock, the crank tooth cocks the firing mechanism and also unlocks the breechblock. After the cocking action, the crank roller presses against the horizontal lower surface of the T slot forcing downward motion of the breechblock.

As the breechblock reaches its lowest limit, the bottom ends of the extractor cams on the breechblock impact the extractor lower lugs, forcing the extractors to rotate anticlockwise. The empty cartridge case, which is gripped by the extractor lips, is levered out of the chamber in one energetic motion. The upper lugs of the extractors now snap over the protrusions of the extractor cams on the breechblock, thereby holding the breechblock in the open position.

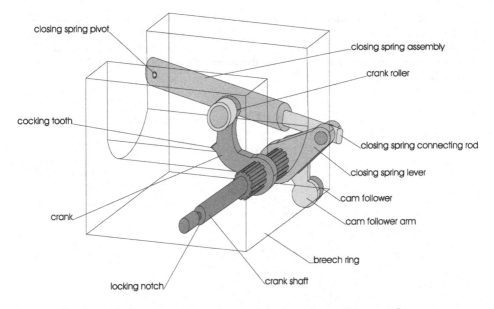

Fig 2.5.1: Components of semi automatic breech operating mechanism

The closing spring gets compressed due to the rotation of the closing spring lever also splined to the crankshaft. The closing spring is retained in the compressed position by the locking of the breechblock by the extractors. During reloading, the breechblock is released by the extractors when the cartridge case rim strikes against the extractor lips.

Components of Semi Automatic Breech Operation

During semi automatic breech operation, three simultaneous operations take place. Firstly the follower splined to the crankshaft causes rotation of the crankshaft about its axis. The crank, which interacts with the breechblock, is also splined to the crankshaft. Hence the crank imparts motion in the vertical plane to the breechblock. The closing spring lever also rotates with the crankshaft causing the closing spring to compress and store energy for the closure of the breech. Basic to all these operations is the interaction of the cam follower with the cam. Hence, the dynamics of semi automatic breech operation is related to the counter recoil linear velocity and acceleration, and the operating cam and follower profile and dimensions. The interrelated angular movements of cam follower arm, closing spring lever and breech operating crank are depicted in Fig 2.5.2. Cams have either of two basic profiles; straight line or circular. Analysis of the dynamics of both cases is investigated ahead in Article 2.6.

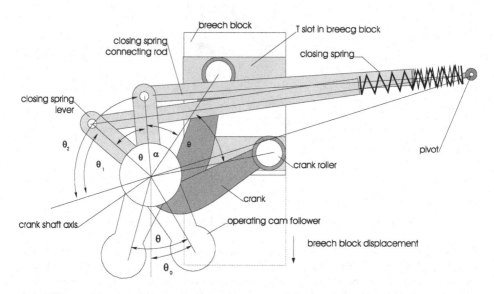

Fig 2.5.2: Crank and closing spring lever rotation as a result of crankshaft rotation due to operating cam-follower action

2.6 Dynamics of Semi Automatic Breech Operation

Operation of Straight Profile Cam

The straight profile cam is defined by the height of the cam and the angle that the slope of the cam makes with the vertical. A typical arrangement is depicted in Fig 2.6.1. The follower of the crankshaft contacts the cam at an initial angle of θ_0 with the vertical. For any subsequent rotation of the crankshaft by an angle θ, the displacement of the crankshaft axis in the direction of counter recoil is given by:

$$s = r_a Sin\theta_0 + (h + r_f)Tan\beta + r_a Sin(\theta - \theta_0) - r_a Cos(\theta - \theta_0)Tan\beta \quad\dots\dots\dots\dots\dots [2.6.1]$$

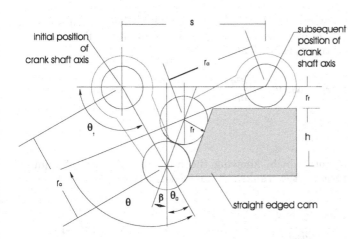

Fig 2.6.1: Diagram showing geometry of straight edged cam operation at initial contact of follower with cam and after crankshaft rotation by an angle θ.

β : cam angle

θ_0 : initial angle of the crankshaft

θ_t : total angular rotation of the crankshaft

θ : instantaneous angular rotation of the crankshaft

r_a: length of cam follower arm

s: displacement of the crankshaft axis in the direction of counter recoil

r_f: cam follower radius

h: effective cam height

Differentiating Equation [2.6.1] with respect to time:

$$\frac{ds}{dt} = r_a Tan\beta Sin(\theta - \theta_0)\frac{d\theta}{dt} + r_a Cos(\theta - \theta_0)\frac{d\theta}{dt}$$

135

Or:

$$\frac{d\theta}{dt} = \omega = \frac{\dfrac{ds}{dt}}{r_a\left[Tan\beta Sin(\theta - \theta_0) + Cos(\theta - \theta_0)\right]} \quad \text{...} [2.6.2]$$

From this equation, the angular speed of the crankshaft can be determined provided the linear velocity of counter recoil is known. Differentiating the equation immediately above again with respect to time:

$$\frac{d^2s}{dt^2} = r_a Tan\beta Cos(\theta - \theta_0)\left(\frac{d\theta}{dt}\right)^2 + r_a Tan\beta Sin(\theta - \theta_0)\frac{d^2\theta}{dt^2}$$

$$-r_a Sin(\theta - \theta_0)\left(\frac{d\theta}{dt}\right)^2 + r_a Cos(\theta - \theta_0)\frac{d^2\theta}{dt^2}$$

This gives an expression from which, knowing the linear acceleration of counter recoil, the angular acceleration of the crankshaft can be determined:

$$\frac{d^2\theta}{dt^2} = \alpha_c = \frac{\dfrac{d^2s}{dt^2} - r_a\omega^2\left[Tan\beta Cos(\theta - \theta_0) - Sin(\theta - \theta_0)\right]}{r_a\left[Tan\beta Sin(\theta - \theta_0) + Cos(\theta - \theta_0)\right]} \quad \text{..................................} [2.6.3]$$

Operation of a Circular Profile Cam

The circular cam consists of a straight vertical portion and a quadrant of a circle. The circular part is defined by the radius of the circle of which it forms a part. In order to express the angular speed and acceleration of rotation of the crankshaft in terms of the linear counter recoil velocity and acceleration, the motion of the follower on the cam is considered in two phases. The first phase begins when the follower contacts the vertical face of the cam and ends when the follower just reaches the curved portion of the cam. This is the beginning of the second phase, which ends when the follower just contacts the horizontal surface of the cam.

Equations of Motion of the Crankshaft during the First Phase

The geometry of a typical semi-automatic breech mechanism is given in Fig 2.6.2, the crankshaft follower is just in contact with the vertical portion of the breech-operating cam. At this instant the shaft has made no rotation about its own axis, which lies perpendicular to the plane of this page.

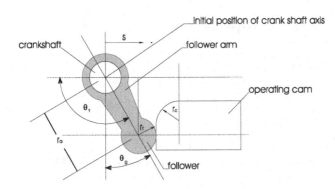

Fig 2.6.3: Position of crankshaft axis at initial contact between follower & breech operating cam.

The displacement of the crankshaft axis in the direction of counter recoil from its position at the instant of initial contact of the follower with the cam, with reference to Figs 2.6.2 and 2.6.3, is given by:

$$s = r_a Sin\theta_0 + r_a Cos(\theta_t - \theta)$$

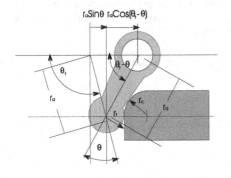

Fig 2.6.3: Fig: Displacement of crankshaft axis during Phase 1.

Since $\theta_t = 90° + \theta_0$:

$$s = r_a Sin\theta_0 + r_a Sin(\theta - \theta_0) ....[2.6.4]$$

Differentiating Equation [2.6.4] with respect to time for the first time:

$$\frac{ds}{dt} = r_a Cos(\theta - \theta_0)\frac{d\theta}{dt}$$

137

From which:

$$\frac{ds}{dt} = \omega = \frac{1}{r_a Cos(\theta - \theta_0)} \frac{ds}{dt} \quad \text{...} [2.6.5]$$

Differentiating Equation [2.6.5] with respect to time:

$$\frac{d^2s}{dt^2} = r_a Cos(\theta - \theta_0)\frac{d^2\theta}{dt^2} - r_a Sin(\theta - \theta_0)\left(\frac{d\theta}{dt}\right)^2$$

From which:

$$\alpha = \frac{\dfrac{d^2s}{dt^2} + r_a Sin(\theta - \theta_0)\left(\dfrac{d\theta}{dt}\right)^2}{r_a Cos(\theta - \theta_0)} \quad \text{.......................................} [2.6.6]$$

Equations of Motion of the Crankshaft during the Second Phase

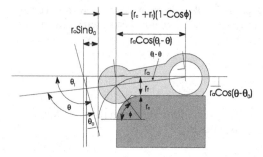

Fig 2.6.4: Displacement of the crankshaft axis during Phase 2

With reference to Fig 2.6.4, when the follower is on the curved portion of the cam surface, its displacement is given by:

$$s = r_a Sin\,\theta_0 + r_a Cos(\theta_t - \theta) + (r_c + r_f) - (r_c + r_f)Cos\phi$$

Or:

$$s = r_a Sin\,\theta_0 + r_a Cos(\theta_t - \theta) + (r_c + r_f)(1 - Cos\phi) \quad \text{..........................} [2.6.7]$$

$$\text{Where: } Cos\phi = \frac{\sqrt{(r_c + r_f)^2 - \{(r_c + r_f) - r_a Cos(\theta - \theta_0)\}^2}}{r_c + r_f}$$

Substituting for $Cos\phi$ in Equation [2.6.7]:

$$s = r_a Sin\theta_0 + r_a Sin(\theta - \theta_0) + r_c + r_f - \sqrt{(r_c + r_f)^2 - [(r_c + r_f) - r_a Cos(\theta - \theta_0)]^2}$$ .... [2.6.8]

$$\therefore r_a Cos(\theta_t - \theta) = r_a Sin(\theta - \theta_0)$$

Differentiating the equation immediately above with respect to time:

$$\frac{ds}{dt} = r_a Cos(\theta - \theta_0)\frac{d\theta}{dt} - \frac{1}{2}\left\{(r_c + r_f)^2 - [(r_c + r_f) - r_a Cos(\theta - \theta_0)]^2\right\}^{-\frac{1}{2}}$$

$$.2r_a Sin(\theta - \theta_0)[(r_c + r_f) - r_a Cos(\theta - \theta_0)]\frac{d\theta}{dt}$$ .... [2.6.9]

Differentiation Equation [2.6.9] with respect to time:

$$\frac{d^2 s}{dt^2} = r_a Cos(\theta - \theta_0)\frac{d\theta}{dt} - r_a Sin(\theta - \theta_0)\left(\frac{d\theta}{dt}\right)^2 - \frac{1}{2}\left\{(r_c + r_f)^2 - [(r_c + r_f) - r_a Cos(\theta - \theta_0)]^2\right\}^{-\frac{1}{2}}$$

$$\cdot\left\{\begin{array}{l}[2[(r_c + r_f) - r_a Cos(\theta - \theta_0)]] \\ \left[r_a Sin(\theta - \theta_0)\frac{d^2\theta}{dt^2} + r_a Cos(\theta - \theta_0)\left(\frac{d\theta}{dt}\right)^2 + 2r_a^2 Sin^2(\theta - \theta_0)\left(\frac{d\theta}{dt}\right)^2\right]\end{array}\right\}$$ [2.6.10]

The angular velocity and acceleration of the crankshaft can be determined from Equations [2.6.9] and [2.6.10].

Displacement of the Breechblock

With reference to Fig 2.6.5, the vertical displacement of the breechblock is related to the angular rotation of the crank about the crank axis by the expression:

$$x = r_i Cos\alpha - r_i Cos(\alpha + \theta)$$ .... [2.6.11]

Differentiating for the first time with respect to time, the vertical speed of the breechblock is:

139

$$\frac{dx}{dt} = r_l Sin(\alpha + \theta)\frac{d\theta}{dt}$$ .... [2.6.12]

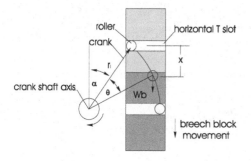

Fig 2.6.5: Rotation of crankshaft and related displacement of breechblock.

Differentiating Equation [2.6.12] with respect to time:

$$\frac{d^2x}{dt^2} = r_l Cos(\alpha + \theta)\left(\frac{d\theta}{dt}\right)^2 - r_l Sin(\alpha + \theta)\frac{d^2\theta}{dt^2}$$ .... [2.6.13]

The moment due to cam action about the crank axis:

$$M_c = m_b r_l Cos\left(\frac{\pi}{2} - \alpha - \theta\right)\frac{d^2x}{dt^2}$$ .... [2.6.14]

Substituting for the acceleration of the breechblock from Equation [2.6.13]:

$$M_c = m_b r_l Cos\left(\frac{\pi}{2} - \alpha - \theta\right)\left[r_l Cos(\alpha + \theta)\left(\frac{d\theta}{dt}\right)^2 - r_l Sin(\alpha + \theta)\frac{d^2\theta}{dt^2}\right]$$ .... [2.6.15]

m_b: mass of breechblock

Displacement of the Closing Spring

From the geometry of Fig 2.6.6, the length of the spring and connecting rod at an angle θ_1 between the line joining the point of attachment of the spring to the breech ring and the centre of the crankshaft and the line through the centre of the closing spring lever is given by:

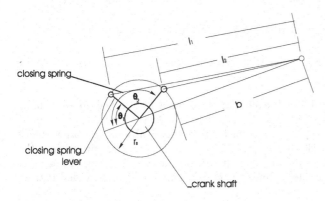

Fig 2.6.6: Relation between crankshaft angle and spring displacement

r_s: closing spring lever length
b: closing spring connecting rod length
l_1: length of closing spring and connecting rod before rotation of crankshaft
l_2: length of closing spring and connecting rod after rotation of crankshaft
θ_1: angle between axis of closing spring & line from pivot to crankshaft axis before rotation
θ_2: angle between axis of closing spring & line from pivot to crankshaft axis after rotation by an angle θ

$$l_1 = \sqrt{\left(r_s Sin\,\theta_1\right)^2 + \left(b + r_s + r_s Cos\,\theta_1\right)^2} \quad\text{...} [2.6.16]$$

The length of the spring and connecting rod when the lever has rotated by an angle θ_2 is given generally by:

141

$$l_2 = \sqrt{(r_s Sin\theta_2)^2 + (b + r_s + r_s Cos\theta_2)^2} \quad \text{...} [2.6.17]$$

Hence the change in length for rotation of the closing spring lever by an angle θ is given by:

$$\delta l = l_1 - l_2 \quad \text{...} [2.6.18]$$

Here $\theta_2 - \theta_1 = \theta_t$ i.e. the total angular rotation of the crankshaft about its axis

The length of the connecting rod is constant so any change in the length is simply the spring displacement.

This condition continues until the breech has descended to the breech fully open position and gets held by the rear lugs of the extractors. At this position the closing spring has to be compressed sufficiently in order to energetically lift the breechblock to the closed position when released by the cartridge case rim striking against the extractor lugs. In other words, the spring torque when the breech is closed must not be less the weight of the breechblock times the crank radius with some added margin of insurance.

2.7 Analysis of Dynamics of Semi-automatic Breech Mechanism

Design Exercise 2.7.1

It is desired to analyze the dynamics of the 76mm semi automatic breech mechanism of known parameters as given below and geometry as depicted in Figs 2.5.1 to 2.6.1.

76 mm Semi-automatic Breech Mechanism Parameters

r_a = 5.0 cm : length of operating cam follower arm
r_f = 1.5 cm: operating cam follower radius
h = 6 cm: cam height; β = 15°: cam surface angle with vertical
x_{max} = 11.76 cm: vertical displacement of breechblock
r_l = 10 cm: crank arm radius
m_b = 30 kg: mass of breechblock
W_b = 392.6 N: weight of breechblock

$\theta_0 = 10°$: Initial angle centre line of the follower with vertical

$\dfrac{ds}{dt} = 2.6$ m/s: velocity of crankshaft in direction of counter recoil

$\dfrac{d^2s}{dt^2} = 1.5$ m/s: linear acceleration of crankshaft in direction of counter recoil

Solution

Total Angular Rotation of Crankshaft

The total angular rotation of the crankshaft is the sum on the idle rotation plus the angular movement necessary to displace the block vertically by the desired 11.76 cm. The idle rotation is dictated by the stipulated displacement of the firing spring and the dimensions of the cocking lever; here the idle rotation is given as 10°.

Vertical displacement of the breechblock is fixed as 11.76 cm. This is 4.16 cm more than the bore radius and gives sufficient clearance for loading the cartridge case taking into consideration cartridge case rim outer diameter. Assuming the direction of counter recoil as the horizontal, from Equation [2.6.11] the angle of rotation of the crankshaft which will result in a vertical displacement of the breechblock of 11.76 cm is 72°. It follows that the initial angle of the crank centre line with the vertical is 90°-72° = 18°.

Hence:

$\alpha = 18°$: initial angle of crank with vertical
$\theta_t = 82°$: total angular rotation of the crankshaft about its axis

Crankshaft Angular Speed

The crankshaft angular speed is determined from Equation [2.6.2]:

$$\omega = \frac{\frac{ds}{dt}}{r_a[Tan\beta Sin(\theta - \theta_0) + Cos(\theta - \theta_0)]}$$. Here the linear speed of the crankshaft axis in

the direction of recoil is given as a constant 2.6 m/s.

Crankshaft Angular Acceleration

The crankshaft acceleration is deduced from Equation [2.6.3]:

$$\alpha_c = \frac{\frac{d^2s}{dt^2} - r_a\omega^2[Tan\beta Cos(\theta - \theta_0) - Sin(\theta - \theta_0)]}{r_a[Tan\beta Sin(\theta - \theta_0) + Cos(\theta - \theta_0)]}$$

Breechblock Acceleration

The breechblock acceleration is determined from Equation [2.6.13].

Torque induced in the Crankshaft by Breechblock Movement

With reference to Fig 2.6.5, the moment arm of the breechblock torque is r_l.Hence the torque induced in the crankshaft vide Equation [2.6.14] is:

$$M_c = m_b\, r_l Cos\left(\frac{\pi}{2} - \alpha - \theta\right)\frac{d^2x}{dt^2}$$

Torque due to Weight of the Breechblock

The torque in the crankshaft due to the weight of the breechblock is:

$$M_b = W_b r_l Cos\left(\frac{\pi}{2} - \alpha - \theta\right)$$

Net Clockwise Torque in Crankshaft

The net clockwise torque is the sum of the torque induced by the motion of the breechblock plus the torque due to the weight of the breechblock. A Matlab programme to compute the net clockwise torque is given next.

Matlab Programme 2.7.1

```
% programme to compute net clockwise torque in crankshaft
% file name: crankshaft torque

mb=30 % breechblock mass kg
ra=.05 % cam follower arm radius m
theta0=10./180*pi % idle angle
theta=linspace(10/180.*pi,82/180*pi,82)
rl=.1 % crank radius m
alfa=18/180*pi % initial crank angle rad
acr=1.5 % counter recoil acceleration m/s^2
vcr=2.6 % counter recoil velocity % m/s
beta=15/180*pi % cam angle rad

g=9.81 % acceleration due to gravity m/s^2
xm=.1176 % maximum breechblock movement m

% Crankshaft angular speed
omega=vcr./(ra.*(tan(beta).*sin(theta-theta0)+cos(theta-theta0)))

% crankshaft angular acceleration
alfac=(acr-ra.*omega.^2.*(tan(beta).*cos(theta-theta0)-sin(theta-
theta0)))./(ra.*(tan(beta).*sin(theta-theta0)+cos(theta-theta0)))

% breechblock acceleration
acb=rl.*cos(alfa+theta-theta0).*omega.^2-rl.*sin(alfa+theta-theta0).*alfac


% crankshaft torque
Mc=mb.*rl.*cos(pi./2-alfa-theta-theta0).*acb
% torque due to weight of breechblock
```

145

Mb=mb.*g.*rl.*cos(pi./2-alfa-theta-theta0)
plot(theta,Mc,theta, Mb)
title('Torques about axis versus angular displacement of crankshaft')
xlabel('Angular displacement radians');ylabel('Torques')
gtext(' torque due to weight of breechblock');gtext('torque due to cam action')

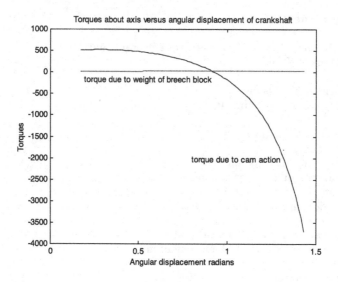

Fig 2.7.1: Torque versus angular displacement curves for 76 mm semi automatic breech mechanism

3

Recoil System Design

3.1 General Principles of Recoil System Design

Functions of the Recoil System

When a gun is fired, the propellant gas force acting on the base of the projectile impels it in the direction of the muzzle while simultaneously propelling the barrel and its attachments backwards. The barrel together with its attachments, which displace during recoil, is called the recoiling parts. The recoiling parts include the barrel, breech mechanism, attachments such as fume extractors and muzzle brakes, dynamic components of the recoil system such as pistons or cylinders and the means to facilitate sliding of the barrel on the cradle.

The function of the recoil system is to limit this rearward movement of the recoiling parts to the least possible distance, as also to return the recoiling parts to the forward or firing position, without adverse effect on the stability of the equipment within the constraint of overall weight of the weapon system and space limitations in the path of the recoiling parts at all angles of elevation and traverse.

With lighter weapon carriages such as of field artillery, stability decides the magnitude of the maximum braking force that can be safely applied to halt the recoiling parts. With heavier weapon carriages like tanks and self propelled artillery pieces, the restricted space in the fighting compartment fixes the recoil length. Large overall weight, hence stability being an inherent feature of such equipment.

Recoil Cycle

On firing, the recoiling parts are propelled rearwards by the propellant gas force acting on the closed breech end of the chamber. This force, varying in magnitude according to the gas pressure inside the barrel, acts during the phase when the projectile is in the muzzle as also after the projectile leaves the muzzle, until such time the pressure in the chamber reduces to atmospheric. This latter phase is commonly called the after effect period. If the barrel is elevated during firing, the weight component of the recoiling parts adds to the gas force in propelling the recoiling parts rearwards.

The recuperator and friction force provide retardation during this initial phase of acceleration of the recoiling parts. Subsequently a hydraulic braking force is introduced gradually to bring the recoiling parts to a halt. At the outset this braking force is minimized to prevent impulse loading of the recoil system components.

As recoil is taking place, energy is being stored in the recuperator. The recuperator serves to return the recoiling parts to the firing position once the rearward movement ceases. This return of the recoiling parts under the influence of the stored energy is called counter recoil, run out or run up.

In guns characterized by a high rate of fire, it is desirable that the recoil cycle be completed in the shortest duration. However, as in recoil, stability consideration and the strength of components involved in the counter recoil phase place a limit on the velocities that can be realized during counter recoil.

Towards the end of counter recoil, the movement is braked additionally by the introduction of further hydraulic throttling or pneumatic braking, in order that the last movement of the recoiling parts be without impact. This additional braking is called buffing and the devices used to implement it are called buffers. Finally, the positioning of impact absorbing material between the parts approaching contact cushions the impact between the moving surfaces of the recoiling parts and the static surfaces of the cradle.

At the end of the recoil-counter recoil strokes, the recoiling parts are maintained in the firing position by the initial recuperator force.

Types of Recoil Systems

Different types of recoil systems exist. The differences lie mainly in the elastic medium of the recuperator, which may be gas or mechanical spring and the interrelation between the recoil brake and the recuperator. The recoil brake and recuperator may be interconnected or separate, but in both cases they function as part of the overall recoil system.

Hydro-spring Recoil Systems

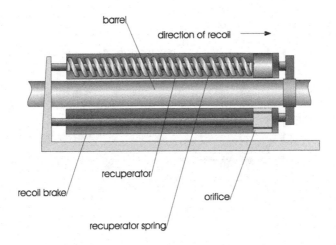

Fig 3.1.1: Hydro-spring recoil system

149

Hydro spring recoil systems are based on hydraulic throttling for recoil braking and counter recoil buffing with mechanical springs employed for recuperation. The hydraulic component of such systems is basically of piston and cylinder type. The arrangement of the recuperator spring varies depending on the amount of energy required to be stored in the spring. This dictates its size. The space available and the configuration of the equipment also influence the spring arrangement. Common arrangements are barrel concentric, recoil brake cylinder concentric and independently located springs.

Hydro-pneumatic Recoil Systems

The hydro-pneumatic recoil system differs from the hydro-spring type in that recuperation is based on energy stored in a gas by compressing it. Dry nitrogen is usually used because of its inert nature. Basically, hydro-pneumatic systems are of either the independent or the dependant type.

Independent Hydro-pneumatic Recoil Systems

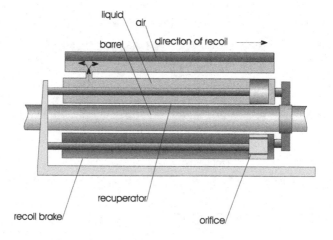

Fig 3.1.2: Independent hydro-pneumatic recoil system

In this arrangement, the recoil brake and the recuperator are separate entities. The piston rods of both the recoil brake and recuperator, or the recoil brake cylinder and the recuperator cylinders, are fixed to the recoiling parts. There is no interaction between the recuperator and the recoil brake.

Dependant Hydro-pneumatic Recoil Systems

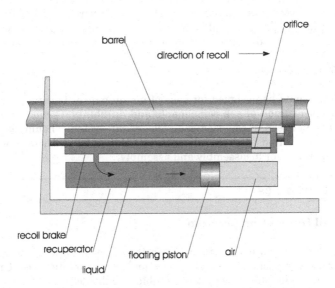

Fig 3.1.3: Dependant hydro-pneumatic recoil system

In the dependant hydro-pneumatic arrangement, only the recoil brake piston rod is attached to the recoiling parts. Oil from the recoil brake is used to compress the gas in the recuperator cylinder with the help of a floating piston.

Buffers

Buffing is achieved by hydraulic throttling or alternatively by pneumatic action. Based on their intended duration of operation buffers may be of two types. The first acts only for a short interval towards the end of the counter recoil stroke. The second type acts over the complete counter recoil stroke.

Selection of a Recoil System Design

A recoil system has to be in keeping with the intended role of the weapon for which it is being designed. Therefore, the individual characteristics of the weapon influence the design of the recoil system. No hard and fast rules exist to spell out the configuration of a recoil system. Factors such as tactical role, rate of fire, elevation range, availability of space, overall weight of the equipment, ground clearance and calibre lead the designer towards final the final arrangement of the recoil system.

Design for Maintenance

A recoil system should be designed for ease of maintenance. This includes simplicity of design, standardization of sub components and durability. Components should be interchangeable between recoil systems. The recoil system should be able to be serviced with a limited degree of technical skill and without the need for sophisticated facilities for disassembly and assembly.

3.2 Recoil Dynamics

Basis of Recoil Design Computation

Recoil design computation is done to establish the exact net braking force necessary to bring the recoiling parts to a halt in the stipulated recoil distance. Once the net braking force has been determined, it is possible to subtract from it the components of recuperator, slide and seal friction forces and arrive at the hydraulic braking force, which is expected of the recoil brake. Secondly, the calculations result in the generation of the velocity of the recoiling parts at successive intervals of displacement during recoil. The hydraulic braking force and the velocity of the recoiling parts at hand, it is possible to arrive at the restriction or the orifice area necessary to generate the desired hydraulic braking force.

Forces Acting During Recoil

The expression for forces acting during recoil is:

$$P - F_B - F_R - \mu(R_1 + R_2) + W_g Sin\phi = m_g \frac{dv_g}{dt} \quad .... \text{Equation [3.4.4] Reference 1}$$

P: propellant gas force
F_B: recoil brake force inclusive of hydraulic & recoil brake seal friction forces
F_R: recuperator force inclusive of the recuperator seal friction force
μ : coefficient of friction at the slides
R_1, R_2: reactions at the supports
W_g: weight of the recoiling parts
ϕ : angle of elevation of the barrel
m_g: mass of the recoiling parts
v_g: velocity of the recoiling parts
t: time

The above equation may be written as:

$$P + W_g Sin\phi - F_{Bnet} = m_g \frac{dx_g^2}{dt^2}$$

$F_{Bnet} = F_B + F_R + \mu(R_1 + R_2)$: net braking force
x_g: displacement of recoiling parts from the firing position at time t.

The accelerating force is hence:

$$F_a = P + W_g Sin\phi - F_{Bnet} \quad .... [3.2.1]$$

Also:

$$F_a = m_g \frac{d^2 x_g}{dt^2}$$

Knowing the accelerating force, the acceleration may be determined from:

$$\frac{d^2 x_g}{dt^2} = \frac{F_a}{m_g} \quad .... [3.2.2]$$

Determination of Approximate Net Recoil Braking Force

In order to bring the recoiling parts to a halt, the work done by the net braking force must equal the energy of free recoil of the recoiling parts plus the work done by the weight component of the recoiling part in moving the recoiling parts through the recoil length. Expressed mathematically:

$$F_{Bnet} l_r = E_g + W_g Sin\phi l_r$$

Or:

$$F_{Bnet} = \frac{E_g}{l_r} + W_g Sin\phi \quad \text{...} [3.2.3]$$

E_g: energy of free recoil of the recoiling parts
l_r: maximum recoil length

The energy of free recoil is given by:

$$E_g = \frac{1}{2} m_g v_{gf}^2$$

v_{gf}: velocity of free recoil

Applying the law of conservation of momentum and using Equation [3.3.1] Reference 1:

$$v_{gf} = \left(\frac{m_p + \xi m_c}{m_g + \xi m_c} \right) v_p$$

m_p: projectile mass
m_c: charge mass
v_p: projectile muzzle velocity
ξ : Serbert factor

Determination of the Precise Net Recoil Braking Force

The value of the net recoil braking force determined vide Equation [2.2.3] is approximate. This value is used initially to establish the exact value. The precise value of the net braking force is found by step-by-step numerical integration over small intervals of time. The procedure is repeated until the selected recoil length is arrived at. The calculations are based on the mass of the recoiling parts and the force-time curve obtained from internal ballistics computation.

Propellant Gas Force-Time Curve

The propellant gas force-time relation is obtained in two phases by internal ballistics computation. The first phase covers the period from shot start to exit. The second or the intermediate ballistics phase is from the instant of projectile exit from the muzzle up to the instant the gas force falls to atmospheric.

Procedure for Determination of Precise Recoil Braking Force

The computation to determine the exact net recoil braking force is carried out in two parts. The first part is from shot start to the end of gas action and the second part is from the end of gas action to the instant when the recoiling parts come to a halt. i.e. when the velocity of recoil is zero.

The time interval δt between successive values of time is taken from the internal ballistics data.

The average gas force P_a for this interval is calculated.

The propelling force, which is the sum of the gas force and the weight component, is determined for the interval. This is given by:

$$F_p = P_a + W_g Sin\phi \quad \text{.. [3.2.4]}$$

The mean accelerating force, in the interval, which is the difference between the propelling force and the net braking force is calculated from:

$$F_a = F_p - F_{Bnet} \quad \text{.. [3.2.5]}$$

The accelerating force divided by the mass of the recoiling parts now gives the acceleration during the small interval in question.

$$a_r = \frac{d^2 x_g}{dt^2} = \frac{F_a}{m_g} \quad \text{...} \quad [3.2.6]$$

During the interval, the change in velocity is given by:

$$\delta v_g = \frac{d^2 x_g}{dt^2} \delta t \quad \text{...} \quad [3.2.7]$$

Starting from the first interval, the velocity at the end of each successive interval can now be calculated:

$$v_{g_n} = v_{g_{n-1}} + \delta v_g \quad \text{...} \quad [3.2.8]$$

From which the mean velocity of each interval is:

$$v_{g_a} = \frac{1}{2}\left(v_{g_{n-1}} + v_{g_n}\right) + v_{g_{n-1}} \quad \text{...} \quad [3.2.9]$$

The displacement in the interval can now be found from:

$$\delta x_g = v_{g_a} \delta t \quad \text{..} \quad [3.2.10]$$

The displacement up to the end of each successive interval becomes:

$$x_{g_n} = x_{g_{n-1}} + \delta x_g \quad \text{..} \quad [3.2.11]$$

Interval from End of Gas Action to End of Recoil

The velocity at the end of this final interval is zero and the time interval for the final interval is calculated from the equation:

$$\delta t_n = \frac{v_{g_{n-1}}}{a_{r_n}} \quad \text{..} \quad [3.2.12]$$

Here n indicates the final or n^{th} interval.

In case the displacement is at variance with the stipulated recoil length, it implies that the value of the net braking force selected was either too high or too low. The net recoil braking force is now corrected by equating the recoil energies according to the equation:

$$F_{Bnet(adjusted)} = \frac{F_{Bnet} \cdot x_{g_n}}{l_r}$$ .... [3.2.13]

The calculation is repeated until the recoil length achieved x is in agreement with the stipulated recoil length l_r.

Cross Check of Results

The results are cross checked by computing the work done in accelerating the recoiling parts during the first part of recoil when the acceleration force is positive and comparing it with the work done in decelerating the recoiling parts during the latter part of the cycle.

The work done in an interval is:

$$\delta E = F_a \delta x$$ .... [3.2.14]

The computation is correct if the positive work done equals the negative work done.

Defining the Components of the Net Braking Force

Determination of Pneumatic Recuperator Force

The initial recuperator force necessary to hold the recoiling parts in the firing position is:

$$F_{R0} = \lambda \left(W_g Sin\phi + \mu W_g Cos\phi + F_{s0} \right)$$ .... Equation [3.4.3] Reference 1

λ : safety factor

F_{S0}: initial friction force of the seals. The initial friction force in the seals is approximated as $0.3F_{R0}$. Hence the expression for initial recuperator force becomes:

$$F_{R0} = \frac{\lambda(W_g Sin\phi + \mu W_g Cos\phi)}{1 - 0.3\lambda} \quad\quad\quad\quad\quad\quad\quad\quad\quad\quad\quad\quad\quad\quad\quad [3.2.15]$$

The initial pressure of gas in the recuperator is hence:

$$p_{R0} = \frac{F_{R0}}{A_R}$$

A_R: area of recuperator piston

Change in volume of the gas by the end of recoil is:

$$\delta V = l_r A_R$$

The volume at the end of recoil is:

$$V_f = V_0 - \delta V$$

V_0: initial volume of the recuperator gas

The recuperator pressure at the end of recoil is related to the initial pressure by the relation:

$$p_{Rf} = \alpha p_{R0}$$

α assumes a value ranging from about 1.5 to about 2 depending on the desired recoil cycle duration which is large for low rate of fire weapons and small for high rate of fire weapons.

From the gas laws:

$$\frac{p_{R0}}{p_{Rf}} = \left(\frac{V_f}{V_0}\right)^n \quad \text{or:} \quad V_0 = \alpha^{\frac{1}{n}} V_f = \alpha^{\frac{1}{n}}(V_0 - \delta V)$$

158

From which:

$$V_0 = \frac{\alpha^{\frac{1}{n}} \delta V}{\alpha^{\frac{1}{n}} - 1}$$ .... [3.2.16]

Applying Equation [3.8.8] Reference 1, again, the pressure in the recuperator at any displacement x_g of the recoiling parts is:

$$p_R = p_{R0} \left(\frac{V_0}{V_0 - A_R x_g} \right)^n$$

The value of V_0 having been established by Equation [3.2.16].

From which, the recuperator force at the same displacement is:

$$F_R = A_R p_R$$ .... [3.2.17]

Determination of Recuperator Seal Friction Force

From the general equation for axial pressure on a seal produced by the seal spring:

$$p_{sp} = \frac{v - K_p}{K_p} p_m$$ .... Equation [3.9.4] Reference 1

p_{sp}: seal spring force
v : Leakage factor of the system
K_p: Pressure factor of the seal
p_m: maximum pressure in the cylinder

Recuperator Seal Friction Force

In the case of the recuperator the maximum pressure occurs at the end of the stroke when the pressure is equal to αp_{R0}, so in this case the seal spring pressure becomes:

$$p_{Rsp} = \frac{\upsilon - K_p}{K_p} \alpha p_{R0}$$

The axial pressure on the recuperator seal is now:

$$p_{Ra} = p_{Rsp} + p_R$$

The radial pressure on the recuperator wall due to the recuperator seal is now:

$$p_{Rr} = K_p p_{Ra}$$

Finally the friction force of the recuperator seal is:

$$F_{Rs} = \mu_R A_{Rs} p_{Rr} \quad \text{..} \quad [3.2.18]$$

A_{Rs}: contact area of the seal on the recuperator wall
μ_R : coefficient of friction at the recuperator seal wall interface

Friction Force of the Slides

Friction of the slides is governed by the weight of the recoiling parts and the displacement of the lines of action of the forces acting on the recoiling parts from the centre of gravity of the recoiling parts which gives rise to moments about the centre of gravity of the recoiling parts. In recoil, the retarding force and the propellant gas force vary; hence the reactions at the slides also vary. However since the propellant gas force duration is small, the reactions at the slides are taken as constant over the entire recoil distance. The friction force of the slides is hence:

$$F_F = \mu(R_1 + R_2)$$

Friction force of the slides is given by:

$$F_F = \mu W_g Cos\phi \dotfill [3.2.19]$$

Computation of Recoil Brake Force

It is now possible to compute the components of the net recoil braking force, defined by Equations [3.2.17], [3.2.18] and [3.2.19] and isolate the hydraulic braking force, which includes the recoil brake seal friction force, from the net braking force. The net braking force is the sum of the hydraulic braking force, the recuperator force and the friction forces of the slides and recuperator seals.

$$F_B = F_{Bnet} - F_F - F_R - F_{Rs} \dotfill [3.2.20]$$

Since the friction force of the recoil brake seal is a function of the hydraulic pressure, it can be computed only after the force of the recoil brake is known in order to obtain the hydraulic recoil brake force.

The pressure in the recoil brake can now be computed from:

$$p_B = \frac{F_B}{A_B} \dotfill [3.2.21]$$

Recoil Brake Seal Friction Force

Having established an expression for pressure in the recoil brake vide Equation [3.2.21], it is possible to compute the recoil brake seal friction force. In the case of the recoil brake, the maximum pressure occurs at the beginning of the stroke and is equal to:

$$p_{B0} = \frac{F_{B0}}{A_B}$$

$F_{B0} = F_{Bnet} - F_{R(0)}$: maximum recoil brake hydraulic force

The spring pressure on the recoil brake seal is hence:

$$p_{Bsp} = \frac{v - K_p}{K_p}\left(\frac{F_{B0}}{A_B}\right)$$

The axial pressure on the recoil brake seal is:

$$p_{Ba} = p_{Bsp} + p_B$$

The radial pressure on the recoil brake wall due to the seal is:

$$p_{Br} = K_p p_{Ba}$$

Finally the friction force of the recoil brake seal is:

$$F_{Bs} = \mu_B A_{Bs} p_{Br} \quad\text{..} \quad [3.2.22]$$

A_{Bs}: contact area of the seal on the recoil brake cylinder wall
μ_B : coefficient of friction at the recoil brake seal wall interface

Hydraulic Braking Force

The hydraulic braking force of the recoil brake is found as follows:

$$F_H = F_B - F_{Bs} \quad\text{..} \quad [3.2.23]$$

Recoil Forces-Displacement Relation

Having determined the net braking force comprised of the recoil brake force, the recuperator force and the friction forces it is possible to see their relation against displacement during recoil as shown in Fig 3.2.1.

Estimation of Recoil Brake Orifice Size

After the seal friction force has been deducted from the recoil brake force what remains is the actual hydraulic braking force. It is now possible to calculate the orifice cross sectional area required for throttling of the fluid in order to generate the necessary hydraulic braking force.

The relation between the hydraulic braking force and the orifice area is given by:

$$F_H = \frac{1}{2a^2} v_g^2 A_B \rho \left(A_B^2 - a^2 \right) \dots\dots\dots\dots\dots\dots\dots\dots\dots\dots \text{ Equation [3.7.8] Reference 1}$$

Since the orifice area is small, in the region of 2% of the piston area, the above equation may be approximated as:

$$F_H = \frac{1}{2a^2} v_g^2 A_B^3 \rho \dots\dots\dots\dots\dots\dots\dots\dots\dots\dots\dots\dots\dots\dots\dots\dots [3.2.24]$$

a: orifice cross sectional area
ρ : mass density of liquid in recoil brake
A_B: piston cross sectional area

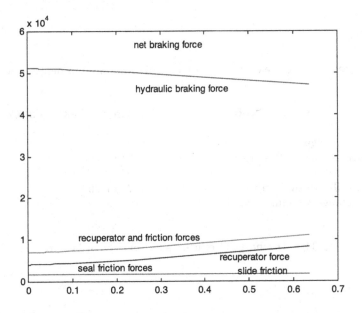

Fig 3.2.1: Graphical representation of recoil forces against recoil displacement

Discharge Coefficient

Also taking into account the viscosity of the fluid and the orifice profile, a factor known as the Discharge Coefficient needs to be applied to the orifice area. Established Discharge Coefficients are tabulated ahead:

The equation immediately above now becomes:

$$F_H = \frac{1}{2C_0^2 a^2} v_g^2 A_B^3 \rho$$

C_0: Discharge Coefficient

From which the orifice area is given by:

$$a = \frac{v_g}{C_0} \sqrt{\frac{A_B^3 \rho}{2F_H}} \quad \text{...} \quad [3.2.25]$$

The recoil travel orifice area relation for a typical recoil brake is depicted in Fig 3.2.3.

Orifice Profile	Discharge Coefficient C_0
Sharp edge	0.6
Rectangular buffer groove	0.71 to 0.83
Control rod	0.95
Recoil throttling valve	0.77 to 0.91
Rectangular counter recoil groove	0.95

Fig 3.2.2: Discharge Coefficients for common orifice profiles

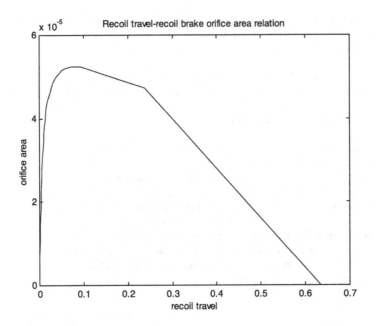

Fig 3.2.3: Recoil travel- recoil brake orifice area relation

3.3 Recoil Computation

Design Exercise 3.3.1

Design of a Tank Gun Recoil System

The recoil system is to be designed for a 100 mm tank gun of muzzle velocity 1000 m/s. Due to space constraints in the fighting compartment, the recoil has to limited to a length of 0.635 m. The elevation range of the gun is from -5° to + 18°. The final recuperator pressure is to be twice the initial. The recoil system is to be of the independent oil and air in contact type.

In the first phase of the design, establish the recoil velocity-travel relation and hence compute the orifice area required to bring the recoiling parts to a halt within the stipulated recoil length.

Necessary internal ballistics data is given in the table at Fig 3.3.1. Schematic diagram of the recoil system along with tentative dimensions is given in Fig 3.3.2.

Symbols Used in the Worksheet

ve: muzzle velocity   mc: charge mass
A: bore cross sectional area   l: shot travel
sigmanetap: ballistic factor   mg: mass of recoiling parts
s: instantaneous shot travel   psi: ballistic factor
delatalam: ballistics factor   p: instantaneous pressure
t: time   P: propellant gas force
pm: maximum propellant gas pressure mp: mass of projectile
tm: time duration from shot start to instant of maximum gas pressure
sm: shot travel from start to instant of maximum gas pressure

lamda: a constant given by: $\lambda = \dfrac{s}{sm}$

Worksheet to plot pressure-time curve						
Calibre	0.1000	l m	2.5870	sigmanetap	0.0875	
ve m/s^	1000.0000	kg	6.2000	pm Pa	2.27E+08	
mc kg	1.0800	kg	402.0000	tm s	0.0017	
A m^2	0.0079			msm m	0.2264	

lamda	s	psi	p	P	deltalam	t
		0	0.00E+00	0.00E+00	0.000	0.00000
0.25	0.057	0.741	1.68E+08	1.32E+06	0.610	0.00103
0.50	0.113	0.912	2.07E+08	1.63E+06	0.780	0.00132
0.75	0.170	0.980	2.22E+08	1.75E+06	0.903	0.00152
1.00	0.226	1.000	2.27E+08	1.78E+06	1.000	0.00169
1.25	0.283	0.989	2.24E+08	1.76E+06	1.081	0.00182
1.50	0.340	0.965	2.19E+08	1.72E+06	1.154	0.00195
1.75	0.396	0.932	2.12E+08	1.66E+06	1.219	0.00206
2.00	0.453	0.898	2.04E+08	1.60E+06	1.282	0.00216
2.50	0.566	0.823	1.87E+08	1.47E+06	1.394	0.00235
3.00	0.679	0.747	1.70E+08	1.33E+06	1.495	0.00252
3.50	0.792	0.675	1.53E+08	1.20E+06	1.589	0.00268
4.00	0.905	0.604	1.37E+08	1.08E+06	1.682	0.00284
4.50	1.019	0.546	1.24E+08	9.73E+05	1.769	0.00298
5.00	1.132	0.495	1.12E+08	8.82E+05	1.851	0.00312
6.00	1.358	0.403	9.15E+07	7.18E+05	2.012	0.00339
7.00	1.585	0.338	7.67E+07	6.02E+05	2.163	0.00365
8.00	1.811	0.284	6.44E+07	5.06E+05	2.309	0.00389
9.00	2.037	0.248	5.63E+07	4.42E+05	2.451	0.00413
10.00	2.264	0.220	4.99E+07	3.92E+05	2.589	0.00437
11.00	2.490	0.199	4.52E+07	3.55E+05	2.725	0.00460
11.42	2.585	0.191	4.34E+07	3.41E+05	2.780	0.00469

Fig 3.3.1: Propellant gas force-time data from shot start to exit

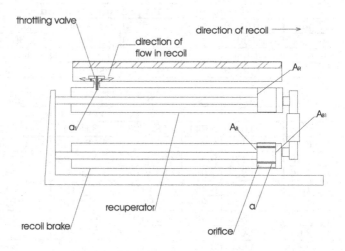

Fig 3.3.2: Physical configuration of proposed recoil system

Recoil System Design Data

Diameter of recoil brake and recuperator pistons: 0.1 m
Diameter of recoil brake and recuperator piston rods: 0.04 m
Initial recuperator force safety factor: 1.15
Coefficient of friction of guides: 0.3
Leakage factor of seals: 1.0
Pressure factor of seal filler: 0.73
Width of recoil brake piston seal: 0.25 m
Width of recuperator piston seal: 0.25 m
Coefficient of friction of seals: 0.05
Discharge Coefficient of recoil orifice: 0.6

Compilation of the Gas Force Time Data

The gas force time data for the first phase i.e. from shot start to exit is extracted from the internal ballistics worksheet. The data for the second phase is computed. Both operations are performed using the following computer programme.

Matlab Programme 3.3.1

```
% programme to compute gas force-time relation
% file name: recoil calculations

format compact
format short e
range1='e8..e29'
P=wk1read('a:recoilcalculations',7,4,range1)
range2='g8..g29'
t=wk1read('a:recoilcalculations',7,6,range2)

pe=4.34.*10.^7
gamma=1.25
ve=1000
mc=1.08
A=pi./4.*.1.^2
se=2.585
a=sqrt(2.*gamma.*pe.*A.*se./((gamma+1).*mc))
p_star=exp(-gamma./2.*(ve./a-a./ve)).*pe
patm=1.013.*10.^5
p1=linspace(pe,p_star,10)
P1=p1.*A
t1=2.*se./(gamma.*(ve+a)).*log(pe./p1)+.00469
p2=linspace(p_star,patm,10)
t2=2.*se./((gamma-1).*a).*(pe./p_star).^.5.*((p_star./p2).^((gamma-1)./(2.*gamma))-1)+.00482
P2=p2.*A
plot(t,P,t1,P1,t2,P2)

title('Propellant gas force-time curve');xlabel('time');ylabel('Gas force')
gtext('projectile exit');gtext('end of gas action')
```

Graphical output of the programme is depicted in Fig 3.3.3.

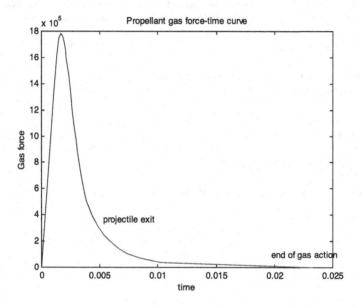

Fig 3.3.3: Propellant gas force-time curve from shot start to end of gas action

Tabulated results are given in Fig 3.3.4 below:

t	P		t	P
0.00	0.00E+00		4.70	3.39E+05
1.03	1.32E+06		4.72	3.36E+05
1.32	1.63E+06		4.73	3.34E+05
1.52	1.75E+06		4.75	3.32E+05
1.69	1.78E+06		4.76	3.30E+05
1.82	1.76E+06		4.78	3.27E+05
1.95	1.72E+06		4.79	3.25E+05
2.06	1.66E+06		4.81	3.23E+05
2.16	1.60E+06		4.82	3.21E+05
2.35	1.47E+06		4.82	3.21E+05
2.52	1.33E+06		5.08	2.85E+05
2.68	1.20E+06		5.39	2.50E+05
2.84	1.08E+06		5.74	2.14E+05
2.98	9.73E+05		6.17	1.78E+05
3.12	8.82E+05		6.70	1.43E+05
3.39	7.18E+05			
3.65	6.02E+05			
3.89	5.06E+05			
4.13	4.42E+05			
4.37	3.92E+05			
4.60	3.55E+05			
4.69	3.41E+05			

Fig 3.3.4: Time-propellant gas force data from shot start to end of gas action.

Determination of Precise Net Braking Force

The precise net braking force is determined as detailed earlier with the help of an Excel workbook, worksheet No 1 of which is given below.

	A	C	D	E	F	G	H	I	J	K	L	M
1	Worksheet to compute recoil velocity, net braking force and displacement: Worksheet No1											
2	mg	402.00	lr	0.64			vgf	16.74				
3	mp	6.20	ve	1000.00			Fnet	89959.59				
4	mc	1.08	S factor	0.50								
5	t	P N	δt	Pa	Fp	Fa=Fp-Fnet	ar	δvg	vg	vga	δxg	xg
6	0.00000	0.00							0.00			0.0000
7			0.00	660350.00	661568.65	571609.06	1421.91	1.42		0.71	0.0007	
8	0.00100	1320700.00							1.42			0.0007
9			0.00	1473100.00	1474318.65	1384359.06	3443.68	1.03		1.94	0.0006	
10	0.00130	1625500.00							2.46			0.0013
11			0.00	1686100.00	1687318.65	1597359.06	3973.53	0.79		2.85	0.0006	
12	0.00150	1746700.00							3.25			0.0019
13			0.00	1764500.00	1765718.65	1675759.06	4168.55	0.83		3.67	0.0007	
14	0.00170	1782300.00							4.08			0.0026
15			0.00	1772500.00	1773718.65	1683759.06	4188.46	0.42		4.29	0.0004	
16	0.00180	1762700.00							4.50			0.0030
17			0.00	1741350.00	1742568.65	1652609.06	4110.97	0.41		4.71	0.0005	
18	0.00190	1720000.00							4.91			0.0035
19			0.00	1690550.00	1691768.65	1601809.06	3984.60	0.80		5.31	0.0011	
20	0.00210	1661100.00							5.71			0.0046
21			0.00	1630800.00	1632018.65	1542059.06	3835.97	0.38		5.90	0.0006	
22	0.00220	1600500.00							6.09			0.0051
23			0.00	1533700.00	1534918.65	1444959.06	3594.43	0.72		6.45	0.0013	
24	0.00240	1466900.00							6.81			0.0064
25			0.00	1399150.00	1400368.65	1310409.06	3259.72	0.33		6.98	0.0007	
26	0.00250	1331400.00							7.14			0.0071
27			0.00	1267250.00	1268468.65	1178509.06	2931.61	0.59		7.43	0.0015	
28	0.00270	1203100.00							7.73			0.0086
29			0.00	1139800.00	1141018.65	1051059.06	2614.57	0.26		7.86	0.0008	
30	0.00280	1076500.00							7.99			0.0094
31			0.00	1024850.00	1026068.65	936109.06	2328.63	0.47		8.22	0.0016	
32	0.00300	973200.00							8.45			0.0111
33			0.00	927750.00	928968.65	839009.06	2087.09	0.21		8.56	0.0009	
34	0.00310	882300.00							8.66			0.0119
35			0.00	800300.00	801518.65	711559.06	1770.05	0.53		8.93	0.0027	
36	0.00340	718300.00							9.19			0.0146
37			0.00	660350.00	661568.65	571609.06	1421.91	0.28		9.33	0.0019	
38	0.00360	602400.00							9.48			0.0165
39			0.00	554300.00	555518.65	465559.06	1158.11	0.35		9.65	0.0029	
40	0.00390	506200.00							9.82			0.0193
41			0.00	474100.00	475318.65	385359.06	958.60	0.19		9.92	0.0020	
42	0.00410	442000.00							10.02			0.0213
43			0.00	417050.00	418268.65	328309.06	816.69	0.25		10.14	0.0030	
44	0.00440	392100.00							10.26			0.0244
45			0.00	373400.00	374618.65	284659.06	708.11	0.14		10.33	0.0021	

Fig 3.3.5: Worksheet to compute recoil velocity, approximate net braking force and recoil displacement

172

	A	C	D	E	F	G	H	I	J	K	L	M
45			0.00	373400.00	374618.65	284659.06	708.11	0.14		10.33	0.0021	
46	0.00460	354700.00							10.40			0.0264
47			0.00	347950.00	349168.65	259209.06	644.80	0.05		10.43	0.0008	
48	0.00468	341200.00							10.45			0.0273
49			0.00	341031.40	342250.05	252290.46	627.59	0.01		10.46	0.0001	
50	0.00469	340862.80							10.46			0.0274
51			0.00	339735.62	340954.26	250994.68	624.36	0.01		10.46	0.0001	
52	0.00470	338608.43							10.47			0.0275
53			0.00	337481.25	338699.90	248740.31	618.76	0.01		10.47	0.0001	
54	0.00472	336354.07							10.48			0.0277
55			0.00	335226.88	336445.53	246485.94	613.15	0.01		10.48	0.0001	
56	0.00473	334099.70							10.49			0.0278
57			0.00	332972.51	334191.16	244231.57	607.54	0.01		10.49	0.0002	
58	0.00475	331845.33							10.49			0.0280
59			0.00	330718.14	331936.79	241977.20	601.93	0.01		10.50	0.0002	
60	0.00476	329590.96							10.50			0.0281
61			0.00	328463.78	329682.42	239722.83	596.33	0.01		10.51	0.0002	
62	0.00478	327336.59							10.51			0.0283
63			0.00	326209.41	327428.05	237468.46	590.72	0.01		10.52	0.0002	
64	0.00479	325082.22							10.52			0.0284
65			0.00	323955.04	325173.68	235214.10	585.11	0.01		10.53	0.0002	
66	0.00481	322827.85							10.53			0.0286
67			0.00	321700.67	322919.32	232959.73	579.50	0.01		10.53	0.0002	
68	0.00482	320573.49							10.54			0.0287
69			0.00	302808.05	304026.69	214067.10	532.51	0.13		10.60	0.0026	
70	0.00506	285042.61							10.67			0.0313
71			0.00	267277.17	268495.82	178536.23	444.12	0.13		10.74	0.0033	
72	0.00537	249511.74							10.80			0.0346
73			0.00	231746.30	232964.94	143005.35	355.73	0.13		10.87	0.0039	
74	0.00572	213980.86							10.93			0.0385
75			0.00	196215.42	197434.07	107474.48	267.35	0.11		10.99	0.0047	
76	0.00615	178449.98							11.04			0.0432
77			0.00	160684.55	161903.19	71943.60	178.96	0.10		11.09	0.0059	
78	0.00668	142919.11							11.14			0.0491
79			0.00	125153.67	126372.32	36412.73	90.58	0.06		11.17	0.0079	
80	0.00739	107388.23							11.20			0.0569
81			0.00	89622.80	90841.44	881.85	2.19	0.00		11.20	0.0115	
82	0.00841	71857.36							11.21			0.0684
83			0.00	54091.92	55310.57	-34649.02	-86.19	-0.16		11.13	0.0204	
84	0.01025	36326.48							11.05			0.0888
85			0.01	18163.24	19381.89	-70577.70	-175.57	-2.25		9.92	0.1273	
86	0.02308	0.00							8.79			0.2161
87			0.04	0.00	1218.65	-88740.94	-220.75	-8.79		4.40	0.1752	
88	0.06292	0.00							0.00			0.3913
89												
90	Adjusted braking force			55438.53								

Fig 3.3.5: Worksheet No 1 (continued)

173

Worksheet No 5 of the workbook is given next. Worksheets No 2 to 4 are identical except that the adjusted braking force of Worksheet No1 replaces the net braking force in Worksheet No 2 and so on. In worksheet No 5 the results are crosschecked by equating the negative and positive work done during recoil vide Equation [3.2.14].

	A	B	C	D	E	F	G	H	I	J	K	L	M	N
1				Worksheet to compute recoil velocity, net braking force and displacement Worksheet No 5										
2	mg		402	lr	0.635	vgf	16.74							
3	mp		6.2	ve	1000	Fnet	58191.72							
4	mc		1.08	S factor	0.5									
5														
6	t		P	δt	Pa	Fp	Fa=Fp-Fnet	ar	δvg	vg	vga	δxg	xg	dE
7	0.00000	0	0.00E+00							0.00			0.0000	
8				0.00100	6.60E+05	6.62E+05	6.03E+05	1.50E+03	1.50E+00		0.75	0.0008		4.53E+02
9	0.00100	1.3207	1.32E+06							1.50			0.0008	
10				0.00030	1.47E+06	1.47E+06	1.42E+06	3.52E+03	1.06E+00		2.03	0.0006		8.62E+02
11	0.00130	1.6255	1.63E+06							2.56			0.0014	
12				0.00020	1.69E+06	1.69E+06	1.63E+06	4.05E+03	8.11E-01		2.96	0.0006		9.65E+02
13	0.00150	1.7467	1.75E+06							3.37			0.0020	
14				0.00020	1.76E+06	1.77E+06	1.71E+06	4.25E+03	8.50E-01		3.79	0.0008		1.30E+03
15	0.00170	1.7823	1.78E+06							4.22			0.0027	
16				0.00010	1.77E+06	1.77E+06	1.72E+06	4.27E+03	4.27E-01		4.43	0.0004		7.60E+02
17	0.00180	1.7627	1.76E+06							4.64			0.0032	
18				0.00010	1.74E+06	1.74E+06	1.68E+06	4.19E+03	4.19E-01		4.85	0.0005		8.18E+02
19	0.00190	1.72	1.72E+06							5.06			0.0036	
20				0.00020	1.69E+06	1.69E+06	1.63E+06	4.06E+03	8.13E-01		5.47	0.0011		1.79E+03
21	0.00210	1.6611	1.66E+06							5.88			0.0047	
22				0.00010	1.63E+06	1.63E+06	1.57E+06	3.91E+03	3.91E-01		6.07	0.0006		9.56E+02
23	0.00220	1.6005	1.60E+06							6.27			0.0053	
24				0.00020	1.53E+06	1.53E+06	1.48E+06	3.67E+03	7.35E-01		6.64	0.0013		1.96E+03
25	0.00240	1.4669	1.47E+06							7.00			0.0067	
26				0.00010	1.40E+06	1.40E+06	1.34E+06	3.34E+03	3.34E-01		7.17	0.0007		9.62E+02
27	0.00250	1.3314	1.33E+06							7.34			0.0074	
28				0.00020	1.27E+06	1.27E+06	1.21E+06	3.01E+03	6.02E-01		7.64	0.0015		1.85E+03
29	0.00270	1.2031	1.20E+06							7.94			0.0089	
30				0.00010	1.14E+06	1.14E+06	1.08E+06	2.69E+03	2.69E-01		8.07	0.0008		8.74E+02
31	0.00280	1.0765	1.08E+06							8.21			0.0097	
32				0.00020	1.02E+06	1.03E+06	9.68E+05	2.41E+03	4.82E-01		8.45	0.0017		1.64E+03
33	0.00300	0.9732	9.73E+05							8.69			0.0114	
34				0.00010	9.28E+05	9.29E+05	8.71E+05	2.17E+03	2.17E-01		8.80	0.0009		7.66E+02
35	0.00310	0.8823	8.82E+05							8.91			0.0123	
36				0.00030	8.00E+05	8.02E+05	7.43E+05	1.85E+03	5.55E-01		9.18	0.0028		2.05E+03
37	0.00340	0.7183	7.18E+05							9.46			0.0150	
38				0.00020	6.60E+05	6.62E+05	6.03E+05	1.50E+03	3.00E-01		9.61	0.0019		1.16E+03
39	0.00360	0.6024	6.02E+05							9.76			0.0170	
40				0.00030	5.54E+05	5.56E+05	4.97E+05	1.24E+03	3.71E-01		9.95	0.0030		1.48E+03
41	0.00390	0.5062	5.06E+05							10.13			0.0199	
42				0.00020	4.74E+05	4.75E+05	4.17E+05	1.04E+03	2.08E-01		10.24	0.0020		8.54E+02
43	0.00410	0.442	4.42E+05							10.34			0.0220	
44				0.00030	4.17E+05	4.18E+05	3.60E+05	8.96E+02	2.69E-01		10.47	0.0031		1.13E+03
45	0.00440	0.3921	3.92E+05							10.61			0.0251	
46				0.00020	3.73E+05	3.75E+05	3.16E+05	7.87E+02	1.57E-01		10.69	0.0021		6.76E+02
47	0.00460	0.3547	3.55E+05							10.77			0.0273	

Fig 3.3.6: Worksheet to compute recoil velocity, precise net braking force and recoil displacement 4th iteration

	A	B	C	D	E	F	G	H	I	J	K	L	M	N
47	0.00460	0.3547	3.55E+05							10.77			0.0273	
48				0.00008	3.48E+05	3.49E+05	2.91E+05	7.24E+02	5.79E-02		10.79	0.0009		2.51E+02
49	0.00468	0.3412	3.41E+05							10.82			0.0281	
50				0.00001	3.41E+05	3.42E+05	2.84E+05	7.07E+02	7.07E-03		10.83	0.0001		3.08E+01
51	0.00469	340862.8029	3.41E+05							10.83			0.0282	
52				0.00001	3.40E+05	3.41E+05	2.83E+05	7.03E+02	9.89E-03		10.84	0.0002		4.31E+01
53	0.00470	338608.4343	3.39E+05							10.84			0.0284	
54				0.00001	3.37E+05	3.39E+05	2.81E+05	6.98E+02	9.88E-03		10.85	0.0002		4.31E+01
55	0.00472	336354.0657	3.36E+05							10.85			0.0286	
56				0.00001	3.35E+05	3.36E+05	2.78E+05	6.92E+02	9.86E-03		10.86	0.0002		4.30E+01
57	0.00473	334099.6971	3.34E+05							10.86			0.0287	
58				0.00001	3.33E+05	3.34E+05	2.76E+05	6.87E+02	9.85E-03		10.87	0.0002		4.30E+01
59	0.00475	331845.3285	3.32E+05							10.87			0.0289	
60				0.00001	3.31E+05	3.32E+05	2.74E+05	6.81E+02	9.83E-03		10.88	0.0002		4.30E+01
61	0.00476	329590.9599	3.30E+05							10.88			0.0290	
62				0.00001	3.28E+05	3.30E+05	2.71E+05	6.75E+02	9.82E-03		10.88	0.0002		4.30E+01
63	0.00478	327336.5913	3.27E+05							10.89			0.0292	
64				0.00001	3.26E+05	3.27E+05	2.69E+05	6.70E+02	9.81E-03		10.89	0.0002		4.29E+01
65	0.00479	325082.2228	3.25E+05							10.90			0.0293	
66				0.00001	3.24E+05	3.25E+05	2.67E+05	6.64E+02	9.79E-03		10.90	0.0002		4.29E+01
67	0.00481	322827.8542	3.23E+05							10.91			0.0295	
68				0.00001	3.22E+05	3.23E+05	2.65E+05	6.59E+02	9.78E-03		10.91	0.0002		4.29E+01
69	0.00482	320573.4856	3.21E+05							10.92			0.0297	
70				0.00024	3.03E+05	3.04E+05	2.46E+05	6.12E+02	1.50E-01		10.99	0.0027		6.61E+02
71	0.00506	285042.6103	2.85E+05							11.07			0.0324	
72				0.00030	2.67E+05	2.68E+05	2.10E+05	5.23E+02	1.59E-01		11.15	0.0034		7.12E+02
73	0.00537	249511.7351	2.50E+05							11.23			0.0357	
74				0.00036	2.32E+05	2.33E+05	1.75E+05	4.35E+02	1.55E-01		11.30	0.0040		7.02E+02
75	0.00572	213980.8598	2.14E+05							11.38			0.0398	
76				0.00043	1.96E+05	1.97E+05	1.39E+05	3.46E+02	1.48E-01		11.46	0.0049		6.82E+02
77	0.00615	178449.9846	1.78E+05							11.53			0.0447	
78				0.00053	1.61E+05	1.62E+05	1.04E+05	2.58E+02	1.38E-01		11.60	0.0062		6.41E+02
79	0.00668	142919.1093	1.43E+05							11.67			0.0508	
80				0.00070	1.25E+05	1.26E+05	6.82E+04	1.70E+02	1.19E-01		11.73	0.0083		5.63E+02
81	0.00739	107388.2341	1.07E+05							11.79			0.0591	
82				0.00102	8.96E+04	9.08E+04	3.26E+04	8.12E+01	8.32E-02		11.83	0.0121		3.96E+02
83	0.00841	71857.35883	7.19E+04							11.87			0.0712	
84				0.00184	5.41E+04	5.53E+04	-2.88E+03	-7.17E+00	-1.32E-02		11.86	0.0218		-6.28E+01
85	0.01025	36326.48359	3.63E+04							11.86			0.0930	
86				0.01283	1.82E+04	1.94E+04	-3.88E+04	-9.65E+01	-1.24E+00		11.24	0.1442		-5.60E+03
87	0.02308	795.6083395	0							10.62			0.2372	
88				0.07492	0.00E+00	1.22E+03	-5.70E+04	-1.42E+02	-1.06E+01		5.31	0.3978		-2.27E+04
89	0.09800		0							0.00			0.6350	
90											dE+	28321.96		
91		Adjusted braking force			58189.13						dE-	-28321.96		

Fig 3.3.6: 4th iteration (continued)

Computation of Recoil Brake Force

The net hydraulic braking force is isolated from the net braking force with the help of the computer programme given below. The steps in the programme, as covered in the previous section, are outlined below:

Displacement of recoiling parts is read from Worksheet No 5.
Net braking force is read from the same worksheet.
Area of recoil brake piston is calculated.
The maximum pressure in the recoil brake is calculated.
The initial recuperator force hence the initial pressure is calculated.
The initial volume of gas in the recuperator is calculated from which the pressure, hence recuperator force, at any displacement is determined.
The recuperator seal friction force is found.
Friction force of the slides is calculated.
The hydraulic braking force including the recoil brake seal friction force is found.
The recoil brake seal friction force is determined and deducted from the hydraulic braking force to give the net hydraulic braking force.
The braking forces are plotted against recoil displacement.

Matlab Programme 3.3.1

```
% programme to compute recoil brake force
% file name: braking force.txt

% displacement of recoiling parts
range1='f4..f45'
xg=wk1read('a:recoil_x_v',3,5,range1)

% maximum pressure in recoil brake
Fnet=58189.13
x=linspace(0,.635,10000)
dBp=.1 % m diameter of recoil brake piston
dBr=.04 % m diameter of recoil brake rod
Ap=pi./4.*(dBp.^2-dBr.^2) % area of recoil brake piston
pBmax=Fnet./Ap

% initial recuperator force
```

drp=.1 % diameter of recuperator piston
drr=.04 % diameter of recuperator piston rod
Ar=pi./4.*(drp.^2-drr.^2) % area of recuperator piston
lamda=1.15 % safety factor
theta= 18./180.*pi % maximum angle of elevation
mug=.3 % coefficient of friction of guides
Wg= 402.*9.81 % weight of recoiling parts

Fr0=lamda.*(Wg.*sin(theta)+mug.*Wg.*cos(theta))./.655

% initial recuperator pressure
pr0=Fr0./Ar

% net change in volume of gas
lr=.635 % total recoil length
dV=Ar.*lr % change in volume at end of stroke

% Final pressure of recuperator
prf = 2.*pr0 % by choice prf=1.5 to 2 pr0

% Initial volume of recuperator
n=1.3
V0=2.^(1./n).*dV/(2.^(1./n)-1)

% gas pressure at displacement xg
pr=pr0.*(V0./(V0-Ar.*xg)).^n
prmax= pr0.*(V0./(V0-Ar.*lr)).^n

% recuperator force
Fr=Ar.*pr
Frmax=Ar.*prmax

% seal friction force of recuperator
Kp=.73 % pressure factor
br=.0127 % length of recuperator piston seal
mus=.05 % coeff of friction of seal
prsp=(1-Kp)./Kp.*prf % spring pressure in recuperator

```
pra=prsp+pr % axial pressure on seals
prr=Kp.*pra % radial pressure of seals on cylinder
Ars=pi.*drp.*br % contact area of recuperator piston
Fsr=mus.* Ars.*prr % friction force
Fsrmax= (prsp+prmax).*Kp.*Ars.*mus
% Slide friction force
Ff=.3.*Wg.*cos(theta)

% recoil brake force
FB= Fnet-Ff-Fsr-Fr
Fsr0=.3.*Fr0
%recoil brake seal friction force
pBmax= (Fnet-Fr0-Ff-Fsr0)./Ap
pBsp=(1-Kp)./Kp.*pBmax % spring pressure in recoil brake
pB=FB/Ap % recoil brake pressure
pBa=pBsp+pB % axial pressure on seals
bB=.0127 % width of recoil brake piston seal
pBr=Kp.*pBa % radial pressure of seals on cylinder
ABs=pi.*dBp.*bB % contact area of recoil brake piston
FsB=mus.* ABs.*pBr % friction force
Fs=Fsr+FsB
FBnet=FB-FsB % hydraulic braking force

% net seal friction force
Fs=FsB+Fsr
Ft=Fs+Fr+Ff % seal +recuperator+sliding friction forces

plot(xg,Fr,x,Fnet,'k',xg,Fs,xg,FBnet,xg,Ft,x,Ff)
title('Recoil system design curves');xlabel('displacement of recoiling parts')
ylabel('braking forces')
gtext('recuperator force');gtext('net braking force');gtext('seal friction
forces');gtext('hydraulic braking force');gtext('recuperator and friction forces')
gtext('slide friction')
```

Graphical Output

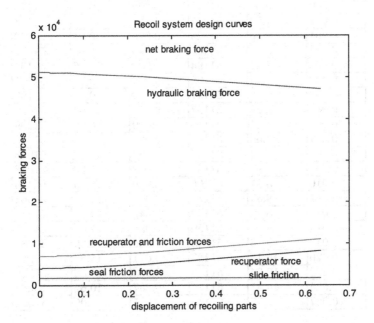

Fig 3.3.7: Plot of recoil forces versus recoil displacement

Graphical output of the computer programme immediately above is shown in Fig 3.3.7.

Determination of Orifice Area

The actual hydraulic braking force found with the help of the computer programme above and velocity and displacement of the recoiling parts extracted from Worksheet No 5 of the recoil calculations are tabulated below in Fig 3.3.10

xg m	vg m/s	FH N	xg m	vg m/s	FH N
0.000	0.00	51203	0.028	10.82	51103
0.001	1.50	51201	0.028	10.83	51103
0.001	2.56	51199	0.028	10.84	51102
0.002	3.37	51196	0.029	10.85	51102
0.003	4.22	51194	0.029	10.86	51101
0.003	4.64	51192	0.029	10.87	51101
0.004	5.06	51191	0.029	10.88	51100
0.005	5.88	51187	0.029	10.89	51099
0.005	6.27	51185	0.029	10.90	51099
0.007	7.00	51180	0.030	10.91	51098
0.007	7.34	51177	0.030	10.92	51098
0.009	7.94	51172	0.032	11.07	51088
0.010	8.21	51169	0.036	11.23	51076
0.011	8.69	51163	0.040	11.38	51061
0.012	8.91	51160	0.045	11.53	51043
0.015	9.46	51150	0.051	11.67	51019
0.017	9.76	51143	0.059	11.79	50988
0.020	10.13	51133	0.071	11.87	50942
0.022	10.34	51125	0.093	11.86	50856
0.025	10.61	51114	0.237	10.62	50202
0.027	10.77	51106	0.635	0.00	47091

Fig 3.3.8: Tabulated values of displacement, velocity of recoiling parts and hydraulic braking force

It is now possible to compute the orifice area with the help of Equation [3.2.25]. The computer programme used is given below and the tabulated output of recoil displacement and orifice area is shown in Fig 3.3.8.

Matlab Programme 3.3.2

```
% programme to compute orifice area of recoil brake
% file name: orifice.txt

A=.0066 % piston area
```

```
C0=.6 % Discharge Coefficient
% recoil displacement
range1='b3..b44'
xg=wk1read('a:recoilorifice',2,1,range1)

% velocity of recoiling parts
range2='c3..c44'
vg=wk1read('a:recoilorifice',2,2,range2)
range3='d3..d44'
Fh=wk1read('a:recoilorifice',2,3,range3) % hydraulic force

row=2.256 % kg/m^3 mass density of liquid
a=vg./C0.*sqrt(A.^3.*row./(2.*Fh))
plot(xg,a)
title('Recoil travel-recoil brake orifice area relation')
xlabel('recoil travel');ylabel('orifice area')
```

xg	a	xg	a
m	m^2	m	m^3
0.000	0.00E+00	0.028	4.78E-01
0.001	6.62E-02	0.028	4.78E-01
0.001	1.13E-01	0.028	4.78E-01
0.002	1.49E-01	0.029	4.79E-01
0.003	1.86E-01	0.029	4.79E-01
0.003	2.05E-01	0.029	4.80E-01
0.004	2.23E-01	0.029	4.80E-01
0.005	2.59E-01	0.029	4.81E-01
0.005	2.76E-01	0.029	4.81E-01
0.007	3.09E-01	0.030	4.81E-01
0.007	3.23E-01	0.030	4.82E-01
0.009	3.50E-01	0.032	4.88E-01
0.010	3.62E-01	0.036	4.96E-01
0.011	3.83E-01	0.040	5.02E-01
0.012	3.93E-01	0.045	5.09E-01
0.015	4.17E-01	0.051	5.15E-01
0.017	4.31E-01	0.059	5.21E-01
0.020	4.47E-01	0.071	5.25E-01
0.022	4.56E-01	0.093	5.24E-01
0.025	4.68E-01	0.237	4.73E-01
0.027	4.75E-01	0.635	0.00E+00

Fig 3.3.9: Tabulated recoil travel-orifice area data

3.4 Dynamics of Counter Recoil

Forces Acting During Counter Recoil

During counter recoil or run out as it is sometimes called, the recuperator force is the propelling force. The retarding forces during this period are that of slide and gland friction and the weight component of the recoiling parts. In the recoil brake, the oil flow is now reversed and can also contribute to the retardation if so desired. In the case of high rate of fire weapons, the retardation due to the reverse flow in the recoil brake may be negated by providing by pass channels for the oil to flow without restriction. In heavy and low rate of fire weapons the retardation by the reverse flow in the recoil brake is supplemented by additional throttling of the liquid in the recuperator or recoil brake. This control over counter recoil velocity may be exercised by arrangements incorporated in either the recoil brake or in the recuperator.

Friction of the Slides

The friction of the slides, as in recoil, is governed by the weight of the recoiling parts and the displacement of the lines of action of the forces acting on the recoiling parts from the centre of gravity of the recoiling parts which gives rise to moments about the centre of gravity of the recoiling parts. This influences the magnitude of the reactions at the slides. In counter recoil, the retarding force and the recuperator force vary; hence the reactions at the slides also vary. However in order to simplify the calculations, the reactions at the slides are taken as constant over the entire counter recoil distance without incurring serious error.

The friction force at the slides vide Equation [3.2.19] is:

$$F_F = \mu(R_1 + R_2)$$

Or:

$$F_F = \mu W_g Cos\phi$$

μ : coefficient of friction between slides and guides
W_g: weight of recoiling parts
φ : angle of elevation

Friction Force of the Seals

The friction force due to the radial force exerted by the recoil brake and recuperator seals on their respective cylinders is a function of the axial pressure inside the cylinders which in turn is related to the velocity of counter recoil. As the two are dependant, it is impossible to estimate the magnitude of one without establishing the magnitude of the other. Hence it is necessary to establish the resistance due to throttling, which when deducted from the recuperator force yields the force on the recuperator and recoil brake pistons, from which the axial pressure on the seals can be deduced. This aspect is dealt with in greater detail further in the text.

Resistance by the Recoil Orifice

Applying Equation [3.2.25], the hydraulic resistance due to the recoil orifice during counter recoil is:

$$F_{Hcr} = \frac{\rho A_{B1}^3}{2C_0^2 a^2} v_{cr}^2$$

A_{B1}: area of the recoil brake piston on the high-pressure side, i.e. area of the piston without deduction of the area of the piston rod.
v_{cr}: velocity of counter recoil

From recoil analysis, the area of the recoil orifice is given by: $a^2 = \dfrac{v_g^2 A_B^3 \rho}{2C_0^2 F_H}$

Hence:

$$F_{Hcr} = \left(\frac{v_{cr}}{v_g}\right)^2 \left(\frac{A_R}{A_{B1}}\right)^3 F_H \dots\dots\dots\dots\dots\dots\dots\dots\dots\dots\dots\dots\dots\dots\dots[3.4.1]$$

The values of v_g and F_H being obtained from recoil computations.

Resistance by the Throttling Orifice

From already established expressions, the resistance provided by the throttling orifice is given by:

$$F_T = \frac{\rho A_R^2}{2C_t^2 a_T^2} v_{cr}^2 \dots\dots\dots\dots\dots\dots\dots\dots\dots\dots\dots\dots\dots\dots\dots\dots[3.4.2]$$

C_t: Discharge Coefficient of the throttling orifice
a_T: area of throttling orifice

Fig 3.4.1 below depicts the effect of throttling on counter recoil velocity. In the case of counter recoil without throttling, the counter recoil velocity increases considerably during the most part of counter recoil, whereas with throttling, the counter recoil velocity is kept fairly constant, which means the acceleration is close to zero.

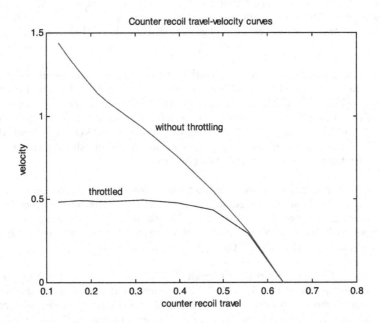

Fig 3.4.1: Counter recoil travel-velocity curves with and without throttling

Equation for the Accelerating Force during Counter Recoil

Seal friction forces neglected for now, the equation for the net counter recoil accelerating force is now:

$$F_{acr} = F_R - F_F - F_S - F_H - F_T = \frac{d^2 x_{cr}}{dt^2} \quad ....[3.4.3]$$

185

Counter Recoil Computations

Counter recoil calculations are carried out at zero elevation angle of the gun. At elevations greater than zero, the weight component of the gun will enhance the retardation so the counter recoil velocities will be less than arrived at after calculation.

The size of the throttling orifice is critical in restricting counter recoil velocities to within acceptable limits. The size of the orifice is fixed purely by trial and error until the selected orifice size yields an acceptable value of counter recoil velocity at the point of buffer contact. If the counter recoil velocity determined is lower than desired, the orifice size is increased and vice versa. It must be kept in mind that in equipments of lesser overall weight, stability considerations during counter recoil limit the magnitude of retardation that can be introduced.

Counter recoil computation follows the method of integration over small successive intervals of counter recoil displacement; the magnitude of the counter recoil velocity is zero at the beginning of the first interval. The counter recoil velocity at the end of the first interval is fixed by a trial and error method. This velocity established then becomes the velocity at the start of the next interval and so on.

The expressions for the resistances offered by the recoil orifice and the throttling orifice given in Equations [3.4.1] and [3.4.2] are modified with the assumption that within the small chosen interval of counter recoil travel, the velocities of recoil and counter recoil and the hydraulic braking force are represented sufficiently accurately by the mean of their values over the small interval.

The expression for resistance offered by the recoil brake in the nth interval therefore becomes:

$$F_{Hcr_n} = \left(\frac{v_{cr_{n-1}} + v_{cr_n}}{v_{g_{n-1}} + v_{g_n}}\right)^2 \left(\frac{A_R}{A_{B1}}\right)^3 \frac{F_{H_{n-1}} + F_{H_n}}{2} \quad \dots \dots \dots \dots \text{[3.4.4]}$$

The subscript n-1 indicates velocity at the start of the nth interval.

Over the small interval of counter recoil displacement, δx_{cr} the expression for resistance offered by the throttling valve becomes:

$$F_{T_n} = \frac{\rho A_R^2}{2C_t^2 a_T^2} \left(\frac{v_{cr_{n-1}} + v_{cr_n}}{2} \right)^2 \quad \text{...................................} [3.4.5]$$

In the small selected interval the acceleration is deemed constant hence the acceleration during the nth interval of counter recoil becomes:

$$a_{cr_n} = \frac{F_{acr_n}}{m_g} \quad \text{...} [3.4.6]$$

The counter recoil velocity v_{cr} determined at the end of each interval is checked with the help of the well known equation:

$$v_{cr_n}^2 = v_{cr_{n-1}}^2 + 2a_{cr_n} \delta x_{cr_n}$$

v_{cr_n} : counter recoil velocity at the end of the nth interval

$v_{cr_{n-1}}$: counter recoil velocity at the beginning of the nth interval

a_{cr_n} r: acceleration of the recoiling parts during the nth interval

δx_{cr_n} : displacement of recoiling parts in the nth interval

Once determined this velocity becomes the counter recoil velocity at the start of the second interval and so on. After the value of the velocity at the end of the interval is fixed, the time duration for the nth interval in question can be calculated from:

$$\delta t_n = \frac{v_{cr_n} - v_{cr_{n-1}}}{a_{cr_n}} \quad \text{...} [3.4.7]$$

The total time elapsed up to the end of the nth interval is:

$$t_{r_n} = t_{n-1} + \delta t_n \quad \text{...} [3.4.8]$$

Friction Force of the Seals

The spring force of the seals, which is calculated, based on the maximum pressure in the recoil brake and recuperator cylinders will remain the same during recoil and counter recoil. However the axial pressure on the seals will be the pressure exerted by the gas in the recuperator transposed according to the piston cross sectional area of the recuperator and the recoil brake pistons.

Iteration one is performed without attempting to take into account the seal friction force of either the recuperator or the recoil brake. As a result of Iteration 1, reasonably accurate values of the force on the recuperator piston are established.

The force on the recuperator piston and the high-pressure side of the recoil brake is:

$$F_{cr} = F_R - F_t \hspace{2cm} [3.4.9]$$

Hence the axial pressure on the recuperator seal is given by:

$$p_{R_a} = \frac{F_{cr}}{A_R} \hspace{2cm} [3.4.10]$$

And the axial pressure on the recoil brake seal is given by:

$$p_{B_a} = \frac{F_{cr}}{A_{B1}} \hspace{2cm} [3.4.11]$$

Friction Force of the Recoil Brake Seal

The recoil brake seal spring force p_{Bsp}, which is the same for recoil, and counter recoil as given by is taken from the recoil calculations.

The axial pressure is given by Equation [3.4.10] above. Hence the recoil brake seal friction force is:

$$F_{Bs} = \mu_s A_{Bs} K_p \left(p_{Bsp} + p_{Ba} \right) \hspace{2cm} 3.4.12]$$

Friction Force of the Recuperator Seal

The recuperator seal spring force p_{Rsp}, which is the same for recoil, and counter recoil as given by Equation [3.9.4] Reference 1, is taken from the recoil calculations.

The friction force on the recuperator cylinder wall is given by:

$$F_{Rs} = \mu_s A_{Rs} K_p \left(p_{Rsp} + p_{Ra} \right) \quad \dotfill \quad [3.4.13]$$

3.5 Counter Recoil Computation

Design Exercise 3.5.1

Counter Recoil Computation

The design calculations are in continuation of the recoil calculations for the same gun as in Article 3.3 of this chapter. The maximum counter recoil velocity is not to exceed 0.85 m/s at the instant of buffer contact. The buffer stroke length is specified as 0.127 m.

The values of recoil displacement, velocity and hydraulic braking force as interpolated from recoil calculations are given in Fig 3.5.1 below:

x_{cr}	x_g	v_g	F_H
0.508	0.127	11.567	50619.00
0.486	0.149	10.410	47504.10
0.464	0.171	9.524	47458.20
0.442	0.193	8.097	47412.30
0.420	0.215	6.940	47366.40
0.398	0.237	5.784	47320.50
0.318	0.317	4.627	47274.60
0.239	0.396	3.470	47228.70
0.159	0.476	2.131	47182.80
0.080	0.555	1.157	47136.90
0.000	0.635	0.000	47091.00

Fig 3.5.1: Values of counter recoil displacement and corresponding recoil displacement, recoil velocity and hydraulic braking force during recoil.

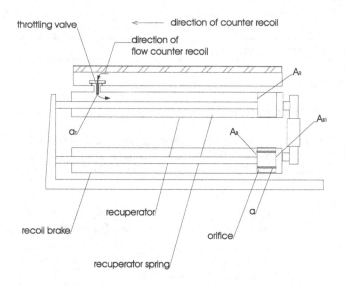

Fig 3.5.2: Schematic layout of the recoil system during counter recoil

Solution

To compute the counter recoil velocity a workbook is created, the first worksheet is constructed as given in Fig 3.5.3. Subsequent worksheets are similar except for the introduction of additional forces, which are explained appropriately. The procedure follows an iterative process as detailed below.

Computation of counter recoil velocity and time upto beginning of buffer stroke													
AB=AR		mg		n	FF		aT		AB1				
pR0		lr		V0	row		Ct		mus				
xcr	xg	dxcr	prxcr	FR	vg	FH	vcr	FHcr	FT	Facr	acr	2acr.dxcr	vcr^2	vcr

Fig 3.5.3: Construction of workbook for counter recoil computation

190

Symbols used in the workbook

AB=AR: areas of recoil brake and recuperator piston less rod area
pR0: initial recuperator pressure
mg: mass of recoiling parts
lr: recoil length
n: gas exponent
V0: initial recuperator volume
FF: slide friction force
row: mass density of liquid
at: throttling orifice area
AB1: recoil brake piston net area
xcr: counter recoil travel
xg: recoil travel
vg: recoil velocity
Ct: Discharge Coefficient of throttling orifice
mus: coefficient of friction of leather seal
dcr: small increment of recoil travel
prxcr: recuperator pressure at counter recoil travel xcr
FR: recuperator force during counter recoil
FH: hydraulic braking force during recoil
FHcr: Recoil brake resistance during counter recoil
Facr: accelerating force during counter recoil
acr: counter recoil acceleration
vcr: velocity of counter recoil

First Calculation

In the first iteration, calculations are carried out without the introduction of the throttling force. The friction force of the seals is also omitted from the calculation at this stage. The first iteration yields the magnitude of the counter recoil velocity when the counter recoil is uncontrolled by throttling.

	B	C	D	E	F	G	H	I	J	K	L	M	N	O	P
1	Computation of counter recoil velocity and time upto beginning of buffer stroke														
	Spreadsheet No1: Counter recoil velocity without throttling, seal friction neglected														
2	AB=AR	6.60E-03	mg	402.00	n	1.30	FF	1183.09	aT	8.25E-06	AB1	8.11E-03			
3	pRO	6.24E+05	lr	0.635	V0	0.01	row	2.26	Ct	6.00E-01	mus	0.05			
4	xcr	xg	dxcr	prxcr	FR	vg	FH	vcr	FHcr	FT	Facr	acr	2acr.dxcr	vcr^2	vcr
5	0.5080	0.1270				11.57	50619.00	1.45							
6			0.0220	7.05E+05	4654.99				1545.09		1926.82	4.79	0.21	2.10	1.45
7	0.4860	0.1490				10.41	50266.20	1.37							
8			0.0220	7.20E+05	4752.55				1704.79		1864.67	4.64	0.20	1.88	1.37
9	0.4639	0.1711				9.25	49913.40	1.29							
10			0.0220	7.35E+05	4853.78				1921.15		1749.55	4.35	0.19	1.67	1.29
11	0.4419	0.1931				8.10	49560.60	1.22							
12			0.0220	7.51E+05	4958.90				2241.54		1534.27	3.82	0.17	1.48	1.22
13	0.4199	0.2151				6.94	49207.80	1.14							
14			0.0220	7.68E+05	5068.11				2768.20		1116.82	2.78	0.12	1.30	1.14
15	0.3978	0.2372				5.78	48855.00	1.09							
16			0.0796	8.09E+05	5336.76				3371.15		782.53	1.95	0.31	1.18	1.09
17	0.3183	0.3167				4.63	48502.20	0.93							
18			0.0796	8.81E+05	5811.74				3866.63		762.03	1.90	0.30	0.87	0.93
19	0.2387	0.3963				3.47	48149.40	0.76							
20			0.0796	9.65E+05	6366.99				4485.21		698.70	1.74	0.28	0.58	0.76
21	0.1591	0.4759				2.31	47796.60	0.55							
22			0.0796	1.06E+06	7023.53				5322.54		517.91	1.29	0.21	0.30	0.55
23	0.0796	0.5554				1.16	47443.80	0.31							
24			0.0796	1.18E+06	7810.15				6390.61		236.45	0.59	0.09	0.09	0.31
25	0.0000	0.6350				0.00	47091.00	0							

Fig 3.5.4: Determination of counter recoil velocity without throttling and neglecting seal friction force

First Iteration

In the second iteration, throttling by means of the throttling orifice is introduced. The throttling orifice area is fixed as a percentage of 0.5 % of the recuperator piston area. The second iteration yields reasonably accurate values of throttling resistance, which when deducted from the recuperator force, gives the force on the recuperator piston and the high pressure side of the recoil brake piston as well. This force divided by the respective piston area yields the pressure in the cylinders necessary for calculation of the seal friction forces. Calculation of the seal friction forces also done in Worksheet No 1 are given ahead.

Computation of counter recoil velocity and time upto beginning of buffer stroke													
Spreadsheet No 2: Counter recoil velocity to establish throttling force													
AB=AR	6.60E-03	n	1.3	FF	1183.086	aT	3.30E-05	AB1	8.11E-03	lr	0.635		
pR0	6.24E+05	V0	0.0101	row	2.256	Ct	6.00E-01	mus	0.05				
xcr	xg	dxcr	prxcr	FR	vg	FH	vcr	FBcr	FT	Facr	acr	2acrdxr	vcr^2	vcr
0.5080	0.1270				11.57	50619.00	1.25							
		0.0220	7.05E+05	4654.99				1159.21	1241.83	1070.87	2.66	0.12	1.56	1.25
0.4860	0.1490				10.41	50266.20	1.20							
		0.0220	7.20E+05	4752.55				1315.45	1136.88	1117.14	2.78	0.12	1.43	1.20
0.4639	0.1711				9.25	49913.40	1.14							
		0.0220	7.35E+05	4853.78				1522.47	1032.69	1115.53	2.77	0.12	1.31	1.14
0.4419	0.1931				8.10	49560.60	1.09							
		0.0220	7.51E+05	4958.90				1816.81	933.60	1025.41	2.55	0.11	1.18	1.09
0.4199	0.2151				6.94	49207.80	1.03							
		0.0220	7.68E+05	5068.11				2290.75	850.85	743.42	1.85	0.08	1.07	1.03
0.3978	0.2372				5.78	48855.00	0.99							
		0.0796	8.09E+05	5336.76				2884.42	725.31	543.94	1.35	0.22	0.98	0.99
0.3183	0.3167				4.63	48502.20	0.88							
		0.0796	8.81E+05	5811.74				3473.44	536.47	618.75	1.54	0.24	0.78	0.88
0.2387	0.3963				3.47	48149.40	0.73							
		0.0796	9.65E+05	6366.99				4229.76	342.20	611.95	1.52	0.24	0.54	0.73
0.1591	0.4759				2.31	47796.60	0.54							
		0.0796	1.06E+06	7023.53				5184.04	162.51	493.90	1.23	0.20	0.29	0.54
0.0796	0.5554				1.16	47443.80	0.31							
		0.0796	1.18E+06	7810.15				6349.76	40.11	237.19	0.59	0.09	0.09	0.31
0.0000	0.6350				0.00	47091.00	0							

Fig 3.5.5: Worksheet to compute throttling force for calculation of seal friction forces

Worksheet to compute seal friction forces							
Kp	0.73	pBmax	7.84E+06	PBsp	2899245.21	mus	0.05
nu	1	pRmax	1.25E+06	pRsp	461404.11	ARs=Abs	0.00405366
FR	**FT**	**Facr**	**pr**	**pB**	**FRs**	**FBs**	**Fs**
4654.990843	1241.83	3413.156843	5.17E+05	4.21E+05	1.45E+02	4.91E+02	6.36E+02
4752.549216	1142.57	3609.979216	5.47E+05	4.45E+05	1.49E+02	4.95E+02	6.44E+02
4853.784223	1038.38	3815.400063	5.78E+05	4.71E+05	1.54E+02	4.99E+02	6.52E+02
4958.897976	934.46	4024.441156	6.10E+05	4.96E+05	1.58E+02	5.02E+02	6.61E+02
5068.107311	856.66	4211.448651	6.38E+05	5.19E+05	1.63E+02	5.06E+02	6.69E+02
5336.761539	736.81	4599.951337	6.97E+05	5.67E+05	1.71E+02	5.13E+02	6.84E+02
5811.738855	546.65	5265.091593	7.98E+05	6.49E+05	1.86E+02	5.25E+02	7.11E+02
6366.993625	346.29	6020.707025	9.12E+05	7.43E+05	2.03E+02	5.39E+02	7.42E+02
7023.533602	163.24	6860.293437	1.04E+06	8.46E+05	2.22E+02	5.54E+02	7.76E+02
7810.145514	40.39	7769.754889	1.18E+06	9.58E+05	2.42E+02	5.71E+02	8.13E+02

Fig 3.5.6: Worksheet to calculate seal friction forces

Symbols used in the Worksheet

pR: pressure on recuperator piston
pB: pressure on recoil brake piston
FRs: recuperator seal friction force during counter recoil
FBs: recoil brake seal friction force during counter recoil
FS: net seal friction force during counter recoil

Second Iteration

The third iteration involves computation of the counter recoil velocity including both the effect of throttling and of seal friction forces. The velocity of counter recoil at the instant of buffer contact is determined with the inclusion of the seal friction forces in the calculation. The orifice area used in the calculations is the same as in the precious iteration. If the counter recoil velocity is in excess of that stipulated at the beginning of the buffer stroke, the orifice area is reduced or vice versa. Successive

194

iterations are performed until the velocity at the beginning of buffer stroke is within acceptable limits.

					Computation of counter recoil velocity upto beginning of buffer stroke										
					Spreadsheet No 4: Counter recoil velocity with throttling and effect of seal friction										
AB=AR	6.60E-03	n		1.3	FF	1183.086	aT	3.30E-05	AB1	8.11E-03	l	0.635			
pR0	6.24E+05	V0		0.0101	row	2.256	Ct	6.00E-01	mus	0.05	ABC=ARC	0.016			
xcr	xg	dxcr	prxcr	FR	vg	FH	vcr	FHcr	FRcr	FRs	Facr	acr	2ardxr	vcr^2	vcr
0.5080	0.1270				11.57	50619.00	1.14								
		0.0220	7.05E+05	4654.99				963.82	1032.53	6.36E+02	839.51	2.09	0.09	1.29	1.14
0.4860	0.1490				10.41	50266.20	1.09								
		0.0220	7.20E+05	4752.55				1094.37	945.81	6.44E+02	885.24	2.20	0.10	1.19	1.09
0.4639	0.1711				9.25	49913.40	1.04								
		0.0220	7.35E+05	4853.78				1267.07	859.45	6.52E+02	891.78	2.22	0.10	1.09	1.04
0.4419	0.1931				8.10	49560.60	0.99								
		0.0220	7.51E+05	4958.90				1512.50	777.23	6.61E+02	825.19	2.05	0.09	0.98	0.99
0.4199	0.2151				6.94	49207.80	0.94								
		0.0220	7.68E+05	5068.11				1904.32	707.32	6.69E+02	604.87	1.50	0.07	0.89	0.94
0.3978	0.2372				5.78	48855.00	0.91								
		0.0796	8.09E+05	5336.76				2413.00	606.77	6.84E+02	449.59	1.12	0.18	0.83	0.91
0.3183	0.3167				4.63	48502.20	0.81								
		0.0796	8.81E+05	5811.74				2955.49	456.47	7.11E+02	505.34	1.26	0.20	0.66	0.81
0.2387	0.3963				3.47	48149.40	0.68								
		0.0796	9.65E+05	6366.99				3633.02	293.92	7.42E+02	514.88	1.28	0.20	0.46	0.68
0.1591	0.4759				2.31	47796.60	0.51								
		0.0796	1.06E+06	7023.53				4485.96	140.62	7.76E+02	437.64	1.09	0.17	0.26	0.51
0.0796	0.5554				1.16	47443.80	0.29								
		0.0796	1.18E+06	7810.15				5564.13	35.14	8.13E+02	214.57	0.53	0.08	0.08	0.29
0.0000	0.6350				0.00	47091.00	0								

Fig 3.5.7: Worksheet to compute counter recoil velocity with throttling and seal friction force included

Third Iteration

The worksheet below shows the counter recoil velocity achieved with an orifice area of 0.25% of the recuperator piston area equal to 16.5 mm², which limits the counter recoil velocity at buffer contact to 0.82 m/s, which is acceptable, the stipulated value being 0.8 m/s.

	B	C	D	E	F	G	H	I	J	K	L	M	N	O	P	Q	R	S
1	Computation of counter recoil velocity up to beginning of buffer stroke																	
	Spreadsheet No 5: Counter recoil velocity and counter recoil time with increased throttling																	
2	AB=AR	6.60E-03	n	1.3	FF	1183.086	aT	1.65E-05	AB1	8.11E-03	lr	0.635						
3	pR0	6.24E+05	V0	0.0101	row	2.256	Ct	6.00E-01	mus	0.05								
4	xcr	xg	dxcr	prxcr	FR	vg	FH	vcr	FHcr	FRcr	FRs	Facr	acr	2acrdxr	vcr^2	vcr	dt	t
5	0.5080	0.1270				11.57	50619.00	0.82										1.31
6			0.0220	7.05E+05	4654.99				506.62	2170.99	6.36E+02	158.26	0.39	0.02	0.67	0.82	0.06	
7	0.4860	0.1490				10.41	50266.20	0.81										1.25
8			0.0220	7.20E+05	4752.55				611.80	2115.09	6.44E+02	198.53	0.49	0.02	0.65	0.81	0.02	
9	0.4639	0.1711				9.25	49913.40	0.79										1.22
10			0.0220	7.35E+05	4853.78				741.24	2011.19	6.52E+02	265.87	0.66	0.03	0.61	0.78	0.04	
11	0.4419	0.1931				8.10	49560.60	0.77										1.18
12	.		0.0220	7.51E+05	4958.90				917.73	1886.43	6.61E+02	310.75	0.77	0.03	0.59	0.77	0.05	
13	0.4199	0.2151				6.94	49207.80	0.75										1.13
14			0.0220	7.68E+05	5068.11				1211.40	1799.86	6.69E+02	205.25	0.51	0.02	0.56	0.75	0.02	
15	0.3978	0.2372				5.78	48855.00	0.73										1.11
16			0.0796	8.09E+05	5336.76				1648.24	1657.92	6.84E+02	163.22	0.41	0.06	0.53	0.73	0.16	
17	0.3183	0.3167				4.63	48502.20	0.69										0.95
18			0.0796	8.81E+05	5811.74				2252.92	1391.89	7.11E+02	272.48	0.68	0.11	0.48	0.69	0.17	
19	0.2387	0.3963				3.47	48149.40	0.61										0.78
20			0.0796	9.65E+05	6366.99				3094.79	1001.56	7.42E+02	345.47	0.86	0.14	0.37	0.61	0.27	
21	0.1591	0.4759				2.31	47796.60	0.48										0.50
22			0.0796	1.06E+06	7023.53				4179.88	524.13	7.76E+02	360.21	0.90	0.14	0.23	0.48	0.16	
23	0.0796	0.5554				1.16	47443.80	0.29										0.34
24			0.0796	1.18E+06	7810.15				5468.91	138.18	8.13E+02	206.75	0.51	0.08	0.08	0.29	0.34	
25	0.0000	0.6350				0.00	47091.00	0										0

Fig 3.5.8: Worksheet to compute counter recoil velocity and time with increased throttling

The output of the various computations is depicted graphically in Fig 3.5.9.

Curves indicated in the figure are as follows:

Curve 1: Counter recoil velocity without throttling. Seal friction forces not included in the computation.

Curve 2: Counter recoil velocity with throttling. Seal friction force not included in calculation.

Curve 3: Counter recoil velocity with throttling, orifice area equal to 0.5 % of recuperator piston area.

Curve 4: Counter recoil velocity with throttling, orifice area equal to 0.25 % of recuperator piston area.

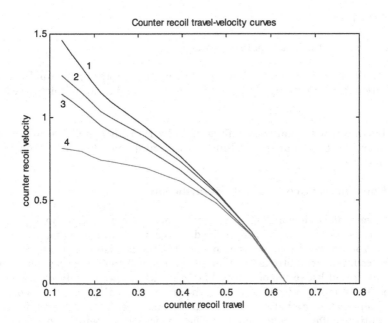

Fig 3.5.9: Graphical depiction of counter recoil travel versus velocity

3.6 Dynamics of the Buffing Phase

Forces Acting during the Buffing Phase

It is desirable during counter recoil to have an energetic return of the recoiling parts to the firing position. It is also necessary that the last movement of the recoiling parts in counter recoil should be without impact on the cradle cap. This dual requirement is achieved by the introduction of a retarding force, in addition to the already present recoil brake orifice retardation and counter recoil throttling, during the last stages of counter recoil movement of the recoiling parts. The phase of counter recoil from introduction of this additional retardation up to the instant when the velocity of the recoiling parts comes to zero is called buffing. The force causing the additional retardation is called the buffing force.

Types of Buffing

Buffing may be effected by one of two means:

Case 1: By the introduction of a buffing force by the throttling of liquid through a constant area orifice resulting in variable retardation and a varying net decelerating force.

Case 2: By the introduction of a buffing force by throttling fluid through a orifice, varying in cross sectional area, resulting in constant deceleration and a constant net decelerating force.

Case 1: Constant Orifice- Variable Retardation

This arrangement is shown diagrammatically in its simplest form in Fig 3.6.1. In essence it consists of a buffer piston fixed to the recoil brake cylinder. The recoil brake piston is hollowed out to accommodate the buffer piston as it approaches the end of the counter recoil stroke. The length over which the buffer piston enters the cavity in the recoil brake piston is the buffer stroke length. The difference between the buffer piston cross sectional area and the cross sectional area of the cavity in the recoil brake piston is the buffer orifice area. In this case the orifice area is constant for the entire buffer stroke length. The orifice area being fixed, with this arrangement, the buffer force is dependant on velocity alone.

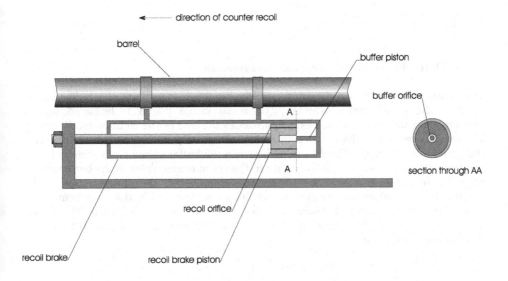

Fig 3.6.1: Buffer with constant orifice variable retardation and variable net decelerating force

The buffer retarding force is given by:

$$F_b = \frac{\rho A_b^2}{2C_b^2 a_b^2} v_{cr}^2 \quad \text{..} [3.6.1]$$

A_b: area of buffer piston
a_b: buffer orifice area
C_b: Discharge Coefficient of the buffer orifice

The equation for the net accelerating force during the buffing phase becomes:

$$F_{acr} = F_R - F_F - F_{Scr} - F_{Bcr} - F_T - F_b \quad \text{...} [3.6.2]$$

The acceleration during this stage is given by:

$$a_{cr} = \frac{F_{acr}}{m_g} \quad .... [3.6.3]$$

Case 2: Variable Orifice- Constant Retardation

The arrangement is shown diagrammatically Fig 3.6.2. It consists of a buffer piston fixed to the recoil brake cylinder. The recoil brake piston is hollowed out to accommodate the buffer piston as it approaches the end of the counter recoil stroke. The length over which the buffer piston enters the cavity in the recoil brake piston is the buffer stroke length. The difference between the buffer piston cross sectional area and the cross sectional area of the cavity in the recoil brake piston is the buffer orifice area. In this case, however, the orifice area varies over the entire buffer stroke length. The variation in orifice area is in accordance with the requirement of producing a constant net decelerating force over the full stroke length. The orifice area being variable, with this arrangement, the buffer force is dependant both on velocity and orifice area.

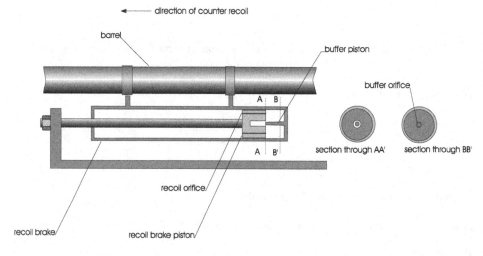

Fig 3.6.2: Variable orifice area, constant retardation and constant net retarding force buffer

In this case the orifice area is designed to vary in order to produce a constant retardation and hence a constant net decelerating force. The accelerating force in this case is:

$$F_{acr} = F_R - F_F - F_{Scr} - F_{Bcr} - F_T$$ .... [3.6.4]

Upon introduction of the buffer force, the net accelerating force becomes:

$$F_k = F_{acr} - F_b$$

Since the acceleration is constant, the net accelerating force is also constant and given by:

$$F_k = m_g a_{cr}$$ .... [3.6.5]

Also, the velocity of counter recoil during the buffer phase is given by:

$$v_{cr}^2 = v_{crb}^2 + 2a_{cr}x_{crb}$$ .... [3.6.6]

a_b: acceleration during buffer stroke
v_{crb}: counter recoil velocity at start of buffer stroke
x_{crb}: buffer stroke at any time t

At the end of the buffer stroke, the counter recoil velocity v_{cr} falls to zero and the acceleration can be determined from:

$$a_{cr} = -\frac{v_{crb}^2}{2S_b}$$ .... [3.6.7]

S_b: buffer stroke length

The buffer force required to maintain this constant net accelerating force is given by:

$$F_b = F_{acr} - F_k$$ .... [3.6.8]

Accordingly the varying buffer orifice area can now be determined from:

201

$$a_b = v_{cr} \sqrt{\frac{\rho A_b^3}{2C_b^2 F_b}}$$ .... [3.6.9]

A representative plot of orifice area against counter recoil travel is depicted in Fig 3.6.3 below:

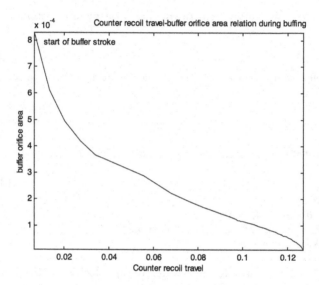

Fig 3.6.3: Counter recoil travel-buffer orifice area relation; variable orifice with constant net retarding force

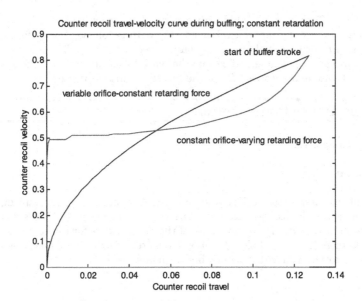

Fig 3.6.4: Counter recoil travel-velocity curves for buffing with constant orifice, varying net retarding force and variable orifice, constant net retarding force during buffing

A comparison of the two means employed for buffing is depicted in Fig 3.6.4. Buffing with a constant orifice and varying net retarding force results in sudden retardation early in the buffing stroke. The minimum velocity is fairly constant for the remainder part of the stroke. As the velocity is low for this part of the stroke, it is impossible to exercise control over the velocity by means of the fixed buffer orifice. Also as seen, the counter recoil velocity, in this case, never falls to zero.

The curve for buffing with variable orifice results in smoother reduction in counter recoil velocity, hence better control over the buffing phase, lower forces at the trunnions and an ultimate value of zero counter recoil velocity. The additional effort required to provide a variable orifice may be worthwhile in terms of the stability of the equipment and near zero velocity of the recoiling parts at impact.

3.7 Buffing Phase Computation

Design Exercise 3.7.1

Buffer Design

Compute the counter recoil velocity during the buffing phase, for buffing with a constant orifice area hence variable retardation. The velocity is 0.815 m/s at the start of buffer stroke of 0.127m, in continuation of the calculations for counter recoil for the same gun as in Article 3.5 of this chapter. Diameter of the buffer piston is 0.0508m and the orifice area given is 0.1% of the buffer piston area.

Solution

The computation is identical to those of counter recoil with the exception of the addition of the retarding force of the buffer. The retarding force of the buffer is given by Equation [3.6.1], suitably modified for application over small selected intervals of buffer stroke:

$$F_b = \frac{\rho A_b^2}{2C_b^2 a_b^2} \frac{v_{cr_{n-1}}^2 + v_{cr_n}^2}{2}$$

$v_{cr_{n-1}}, v_{cr_n}$: counter recoil velocities at the beginning and end of the nth interval under consideration

Additional Symbols used in the Worksheet

Cb: Discharge Coefficient for buffer orifice
Ab: area of buffer piston
ab: buffer orifice area
Fb: buffer force. Fscr: seal friction force during counter recoil

The worksheet is given in Fig 3.7.1 below:

	C	D	E	F	G	H	I	J	K	L	M	N	O	P	Q	R
1	Worksheet for computation of counter recoil velocity & time during buffing phase															
	Constant buffer orifice area with variable deceleration															
2	AB+AR	0.0066	n	1.3	FF	1125.24	aT	8.25E-06	AB1	0.00810732	Ab	2.03E-03				
3	pR0	623750	V0	0.0101	row	2.26	Ct	6.00E-01	Cb	0.6	ab	2.03E-06				
4	xg	dxg	FR	vg	FH	vcr	FBcr	FT	Fb	FScr	Facr	acr	vcr^2	vcr	dt	tcr
5	0.0000			0.00	46374	0.36										
6		0.0008	4118.06				11831.21	2052.08	984.66	557.46	-12432.60	-30.93	0.13	0.36	0.002	0.248
7	0.0008			1.50	46371	0.42						0.00				
8		0.0006	4120.44				3580.15	872.77	1163.29	618.84	-3239.84	-8.06	0.18	0.42	0.001	0.246
9	0.0014			2.56	46369	0.43						0.00				
10		0.0006	4122.55				1829.75	907.23	1209.23	617.61	-1566.51	-3.90	0.19	0.43	0.001	0.245
11	0.0020			3.37	46367	0.44						0.00				
12		0.0008	4124.92				1147.37	926.66	1235.12	617.70	-927.18	-2.31	0.19	0.44	0.002	0.244
13	0.0027			4.22	46365	0.44						0.00				
14		0.0004	4127.03				860.24	938.66	1251.12	617.79	-666.02	-1.66	0.20	0.44	0.001	0.242
15	0.0032			4.64	46364	0.44						0.00				
16		0.0005	4128.66				722.37	945.44	1260.15	617.85	-542.39	-1.35	0.20	0.44	0.001	0.241
17	0.0036			5.06	46362	0.45						0.00				
18		0.0011	4131.44				571.86	953.95	1271.50	617.97	-409.08	-1.02	0.20	0.45	0.002	0.240
19	0.0047			5.88	46358	0.45						0.00				
20		0.0006	4134.43				469.87	961.63	1281.74	618.09	-322.14	-0.80	0.20	0.45	0.001	0.237
21	0.0053			6.27	46356	0.45						0.00				
22		0.0013	4137.84				395.31	968.07	1290.32	618.23	-259.33	-0.65	0.20	0.45	0.003	0.236
23	0.0067			7.00	46352	0.45						0.00				
24		0.0007	4141.45				341.51	974.05	1298.29	618.37	-216.01	-0.54	0.20	0.45	0.002	0.233
25	0.0074			7.34	46350	0.45						0.00				
26		0.0015	4145.42				302.21	979.22	1305.19	618.54	-184.98	-0.46	0.20	0.45	0.003	0.231
27	0.0089			7.94	46345	0.45						0.00				
28		0.0008	4149.56				272.12	984.02	1311.58	618.61	-162.02	-0.40	0.21	0.45	0.002	0.228
29	0.0097			8.21	46342	0.45						0.00				
30		0.0017	4153.99				249.41	988.36	1317.37	617.72	-144.11	-0.36	0.21	0.45	0.004	0.226
31	0.0114			8.69	46337	0.46						0.00				
32		0.0009	4158.56				231.10	992.49	1322.86	616.45	-129.59	-0.32	0.21	0.46	0.002	0.223
33	0.0123			8.91	46334	0.46						0.00				
34		0.0028	4165.04				212.98	997.50	1329.54	616.22	-116.45	-0.29	0.21	0.46	0.006	0.221
35	0.0150			9.46	46325	0.46						0.00				
36		0.0019	4173.40				195.75	1003.61	1337.69	616.56	-105.45	-0.26	0.21	0.46	0.004	0.215
37	0.0170			9.76	46319	0.46						0.00				
38		0.0030	4182.21				183.79	1009.52	1345.57	616.92	-98.83	-0.25	0.21	0.46	0.007	0.211
39	0.0199			10.13	46309	0.46						0.00				
40		0.0020	4191.27				174.53	1015.23	1353.18	617.29	-94.20	-0.23	0.21	0.46	0.004	0.204
41	0.0220			10.34	46302	0.46						0.00				
42		0.0031	4200.66				167.55	1020.74	1360.52	617.67	-91.07	-0.23	0.21	0.46	0.007	0.200
43	0.0251			10.61	46292	0.46						0.00				
44		0.0021	4210.24				161.80	1026.26	1367.88	618.06	-88.99	-0.22	0.21	0.46	0.005	0.193
45	0.0273			10.77	46285	0.46						0.00				
46		0.0009	4215.71				159.05	1029.36	1372.01	618.28	-88.23	-0.22	0.22	0.46	0.002	0.189
47	0.0281			10.82	46282	0.47						0.00				

Fig 3.7.1: Worksheet for computation of counter recoil velocity during buffing phase;
Constant orifice area with variable deceleration.

	C	D	E	F	G	H	I	J	K	L	M	N	O	P	Q	R
47	0.0281			10.82	46282	0.47							0.00			
48		0.0001	4217.48				158.26	1030.46	1373.48	618.35	-88.32	-0.22	0.22	0.47	0.000	0.187
49	0.0282			10.83	46281	0.47							0.00			
50		0.0002	4217.96				158.08	1030.91	1374.07	618.37	-88.72	-0.22	0.22	0.47	0.000	0.187
51	0.0284			10.84	46281	0.47							0.00			
52		0.0002	4218.52				157.86	1031.35	1374.66	618.40	-88.99	-0.22	0.22	0.47	0.000	0.186
53	0.0286			10.85	46280	0.47							0.00			
54		0.0002	4219.08				157.64	1031.79	1375.26	618.42	-89.27	-0.22	0.22	0.47	0.000	0.186
55	0.0287			10.86	46280	0.47							0.00			
56		0.0002	4219.65				157.42	1032.24	1375.85	618.44	-89.54	-0.22	0.22	0.47	0.000	0.185
57	0.0289			10.87	46279	0.47							0.00			
58		0.0002	4220.22				157.20	1032.68	1376.44	618.47	-89.81	-0.22	0.22	0.47	0.000	0.185
59	0.0290			10.88	46279	0.47							0.00			
60		0.0002	4220.79				156.98	1033.12	1377.03	618.49	-90.07	-0.22	0.22	0.47	0.000	0.184
61	0.0292			10.89	46278	0.47							0.00			
62		0.0002	4221.37				156.77	1033.57	1377.62	618.51	-90.33	-0.22	0.22	0.47	0.000	0.184
63	0.0293			10.90	46278	0.47							0.00			
64		0.0002	4221.96				156.55	1034.01	1378.21	618.54	-90.59	-0.23	0.22	0.47	0.000	0.183
65	0.0295			10.91	46277	0.47							0.00			
66		0.0002	4222.55				156.34	1034.46	1378.80	618.56	-90.84	-0.23	0.22	0.47	0.000	0.183
67	0.0297			10.92	46277	0.47							0.00			
68		0.0027	4227.77				154.42	1036.90	1382.06	618.38	-89.23	-0.22	0.22	0.47	0.005	0.183
69	0.0324			11.07	46268	0.47							0.00			
70		0.0034	4238.93				150.95	1042.47	1389.48	618.33	-87.53	-0.22	0.22	0.47	0.007	0.178
71	0.0357			11.23	46256	0.47							0.00			
72		0.0040	4252.61				148.01	1051.40	1401.39	618.79	-92.23	-0.23	0.22	0.47	0.011	0.171
73	0.0398			11.38	46242	0.47							0.00			
74		0.0049	4269.17				145.60	1062.51	1416.20	618.96	-99.34	-0.25	0.22	0.47	0.010	0.160
75	0.0447			11.53	46226	0.47							0.00			
76		0.0062	4289.92				143.79	1076.06	1434.26	618.80	-108.23	-0.27	0.22	0.47	0.013	0.150
77	0.0508			11.67	46205	0.48							0.00			
78		0.0083	4317.24				143.25	1096.44	1461.43	618.48	-127.60	-0.32	0.23	0.48	0.017	0.137
79	0.0591			11.79	46176	0.48							0.00			
80		0.0121	4356.30				145.59	1134.59	1512.27	618.10	-179.50	-0.45	0.23	0.48	0.025	0.120
81	0.0712			11.87	46133	0.49							0.00			
82		0.0218	4422.70				163.40	1282.63	1709.59	614.70	-472.87	-1.18	0.24	0.49	0.042	0.095
83	0.0930			11.86	46054	0.54							0.00			
84		0.0068	4480.09				188.58	1476.45	1967.93	609.66	-887.77	-2.21	0.29	0.54	0.012	0.053
85	0.0998			11.83	46040	0.57							0.00			
86		0.0068	4507.87				211.96	1651.79	2201.64	604.83	-1287.59	-3.20	0.32	0.57	0.012	0.041
87	0.1066			11.80	46025	0.61							0.00			
88		0.0068	4535.95				245.74	1906.10	2540.60	597.22	-1878.96	-4.67	0.37	0.61	0.011	0.029
89	0.1134			11.77	46011	0.66							0.00			
90		0.0068	4564.35				295.40	2280.55	3039.70	585.01	-2761.55	-6.87	0.43	0.66	0.010	0.019
91	0.1202			11.74	45996	0.73							0.00			
92		0.0068	4593.06				368.90	2834.62	3778.21	566.62	-4080.54	-10.15	0.53	0.73	0.009	0.009
93	0.1270			11.71	45982	0.82										

Fig 3.7.1(Continued): Worksheet for computation of counter recoil velocity during buffing phase; Constant orifice area with variable deceleration.

	T	U	V	W	X	Y	Z	AA
1			Worksheet to compute seal friction force					
2	Kp	0.73	pB0	7.84E+06	PBsp	2.90E+06	mus	0.05
3	nu	1		1.25E+06	pRsp	4.61E+05	Abs=Ars	4.05E-03
4	FR	FT	Facr	pR	pB	FRScr	FBScr	FScr
5	4118.06	2637.11	1480.95	224386.22	182668.14	101.47	456.00	557.46
6	4120.44	1130.27	2990.18	453056.98	368824.25	135.30	483.54	618.84
7	4122.55	1162.75	2959.80	448453.88	365076.96	134.62	482.98	617.61
8	4124.92	1162.75	2962.17	448812.94	365369.26	134.67	483.03	617.70
9	4127.03	1162.75	2964.28	449132.58	365629.48	134.72	483.07	617.79
10	4128.66	1162.75	2965.91	449379.76	365830.70	134.76	483.10	617.85
11	4131.44	1162.75	2968.68	449800.59	366173.29	134.82	483.15	617.97
12	4134.43	1162.75	2971.68	450254.44	366542.76	134.89	483.20	618.09
13	4137.84	1162.75	2975.09	450771.17	366963.41	134.96	483.26	618.23
14	4141.45	1162.75	2978.70	451318.03	367408.61	135.05	483.33	618.37
15	4145.42	1162.75	2982.67	451919.50	367898.25	135.13	483.40	618.54
16	4149.56	1165.11	2984.45	452189.28	368117.87	135.17	483.43	618.61
17	4153.99	1191.29	2962.70	448894.24	365435.45	134.69	483.04	617.72
18	4158.56	1227.20	2931.36	444145.13	361569.30	133.98	482.47	616.45
19	4165.04	1239.29	2925.75	443294.97	360877.20	133.86	482.36	616.22
20	4173.40	1239.29	2934.11	444562.21	361908.83	134.05	482.52	616.56
21	4182.21	1239.29	2942.92	445896.32	362994.90	134.24	482.68	616.92
22	4191.27	1239.29	2951.98	447269.61	364112.87	134.45	482.84	617.29
23	4200.66	1239.29	2961.36	448691.61	365270.49	134.66	483.01	617.67
24	4210.24	1239.29	2970.95	450144.13	366452.95	134.87	483.19	618.06
25	4215.71	1239.29	2976.42	450972.37	367127.21	134.99	483.29	618.28
26	4217.48	1239.29	2978.19	451241.00	367345.90	135.03	483.32	618.35
27	4217.96	1239.29	2978.67	451313.07	367404.57	135.04	483.33	618.37
28	4218.52	1239.29	2979.22	451397.67	367473.44	135.06	483.34	618.40
29	4219.08	1239.29	2979.79	451482.93	367542.85	135.07	483.35	618.42
30	4219.65	1239.29	2980.35	451568.86	367612.80	135.08	483.36	618.44
31	4220.22	1239.29	2980.93	451655.48	367683.32	135.10	483.37	618.47
32	4220.79	1239.29	2981.50	451742.79	367754.39	135.11	483.38	618.49
33	4221.37	1239.29	2982.08	451830.80	367826.04	135.12	483.39	618.51
34	4221.96	1239.29	2982.67	451919.53	367898.27	135.13	483.40	618.54
35	4222.55	1239.29	2983.26	452008.97	367971.08	135.15	483.41	618.56
36	4227.77	1249.05	2978.72	451321.56	367411.44	135.05	483.33	618.38
37	4238.93	1261.26	2977.67	451162.73	367282.18	135.02	483.31	618.33
38	4252.61	1263.71	2988.90	452863.26	368666.54	135.27	483.52	618.79
39	4269.17	1276.04	2993.14	453505.38	369189.28	135.37	483.59	618.96
40	4289.92	1300.82	2989.11	452895.36	368692.67	135.28	483.52	618.80
41	4317.24	1335.97	2981.27	451707.52	367725.68	135.10	483.38	618.48
42	4356.30	1384.36	2971.94	450294.25	366575.17	134.89	483.21	618.10
43	4422.70	1534.31	2888.39	437634.21	356268.89	133.02	481.68	614.70
44	4480.09	1715.76	2764.33	418838.05	340967.33	130.24	479.42	609.66
45	4507.87	1862.27	2645.59	400847.71	326321.76	127.58	477.25	604.83
46	4535.95	2077.40	2458.55	372508.10	303251.08	123.38	473.84	597.22
47	4564.35	2406.16	2158.19	326997.83	266202.12	116.65	468.36	585.01
48	4593.06	2886.97	1706.08	258497.16	210437.15	106.52	460.10	566.62

Fig 3.7.2: Worksheet to compute seal friction forces during buffing

Graphical representation of forces during buffing phase of counter recoil in the case of constant orifice with variable retardation is shown for design problem just solved.

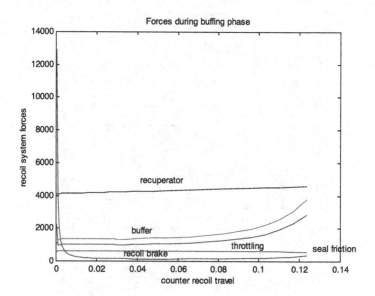

Fig 3.7.3: Recoil system forces during buffing phase of counter recoil with constant orifice

Computation of Orifice Area during Buffing

Compute the counter recoil velocity and the orifice area during the buffing phase, for buffing with variable orifice and constant deceleration. The velocity is 0.815 m/s at the start of buffer stroke of 0.127m in continuation of the calculations for counter recoil for the same gun as in Section 5 of this chapter. Diameter of the buffer piston is 0.0381m. Discharge Coefficient of the buffer orifice is 0.6.

Solution

The acceleration during the buffing phase is given by Equation [3.6.7]:

$$a_{cr} = -\frac{0.815^2}{2x0.127} = -2.615 \text{ m/s}^2$$

The constant net decelerating force is given by Equation [3.6.5]:

$$F_k = 402(-2.615) = -1051.23 \text{ N}$$

The calculations to compute counter recoil velocity with the help of Equation [3.6.6] and buffer force with Equation [3.6.8] are shown in the following worksheet.

Friction force of the seals is computed as in the earlier exercise.

Finally with the help of Equation [3.6.9] the buffer orifice is computed.

	B	C	D	E	F	G	H	I	J	K	L	M	N	O	P
1	Worksheet to compute velocity and buffer orifice area during buffing phase														
1	Variable buffer orifice area with constant deceleration														
2	AB	0.0066	n			1.3	FF	1125.239862	aT	8.25E-06	AB1	8.11E-03	Fk	-1051.25374	Asr=AsB	0.00405366
3	pR0	623750	V0			0.0101	row	2.258	Ct	6.00E-01	acr	-2.62E+00	Db		0.0381	prsp	461404.1096
4	Kp	0.73	nu			1	pB0	7.84E+06	prmax	1.25E+06	mus		0.05	Cb		0.6	PBsp	2899245.205
5	xg	dxg	xb	prxcr	FR	vg	FH	ver	FBcr	FT	Facr	Fk	Fb	FScr	ab
6	0.0000					0.00	51203	0.00							
7		0.0008	0.1270	623948.88	4118.06				165.35	25.97	1795.18	-1051.25	2846.43	1006.32	8.77E-06
8	0.0008					1.50	51201	0.06							
9		0.0006	0.1262	624309.40	4120.44				119.06	26.29	2849.85	-1051.25	3901.10	1006.50	1.76E-05
10	0.0014					2.56	51199	0.08							
11		0.0006	0.1256	624628.35	4122.55				91.87	41.26	2864.18	-1051.25	3915.43	1005.36	2.21E-05
12	0.0020					3.37	51196	0.10							
13		0.0008	0.1250	624987.41	4124.92				79.42	58.09	2862.16	-1051.25	3913.42	1004.08	2.63E-05
14	0.0027					4.22	51194	0.12							
15		0.0004	0.1243	625307.05	4127.03				73.94	73.07	2854.79	-1051.25	3906.04	1002.94	2.96E-05
16	0.0032					4.64	51192	0.13							
17		0.0005	0.1238	625554.23	4128.66				71.40	84.63	2847.38	-1051.25	3898.64	1002.06	3.19E-05
18	0.0036					5.06	51191	0.14							
19		0.0011	0.1234	625975.06	4131.44				69.05	104.31	2832.84	-1051.25	3884.09	1000.56	3.54E-05
20	0.0047					5.88	51187	0.16							
21		0.0006	0.1223	626428.91	4134.43				67.71	125.51	2815.97	-1051.25	3867.22	998.94	3.90E-05
22	0.0053					6.27	51185	0.17							
23		0.0013	0.1217	626945.64	4137.84				67.46	149.61	2795.53	-1051.25	3846.79	997.11	4.26E-05
24	0.0067					7.00	51180	0.19							
25		0.0007	0.1203	627492.50	4141.45				67.78	175.08	2773.35	-1051.25	3824.61	995.17	4.63E-05
26	0.0074					7.34	51177	0.20							
27		0.0015	0.1196	628093.97	4145.42				69.19	203.04	2747.95	-1051.25	3799.20	993.04	5.00E-05
28	0.0089					7.94	51172	0.22							
29		0.0008	0.1181	628720.74	4149.56				70.88	232.13	2721.30	-1051.25	3772.56	990.83	5.37E-05
30	0.0097					8.21	51169	0.23							
31		0.0017	0.1173	629392.29	4153.99				73.35	263.25	2692.15	-1051.25	3743.41	988.46	5.74E-05
32	0.0114					8.69	51163	0.24							
33		0.0009	0.1156	630084.65	4158.56				75.91	295.26	2662.14	-1051.25	3713.40	986.03	6.10E-05
34	0.0123					8.91	51160	0.25							
35		0.0028	0.1147	631066.36	4165.04				80.29	340.55	2618.96	-1051.25	3670.21	982.59	6.59E-05
36	0.0150					9.46	51150	0.28							
37		0.0019	0.1120	632333.60	4173.40				85.89	398.83	2563.44	-1051.25	3614.70	978.16	7.19E-05
38	0.0170					9.76	51143	0.30							
39		0.0030	0.1100	633667.71	4182.21				92.46	459.96	2504.55	-1051.25	3555.81	973.52	7.78E-05
40	0.0199					10.13	51133	0.32							
41		0.0020	0.1071	635041.00	4191.27				99.21	522.64	2444.18	-1051.25	3495.43	968.77	8.37E-05
42	0.0220					10.34	51125	0.34							
43		0.0031	0.1050	636463.00	4200.66				106.45	587.30	2381.67	-1051.25	3432.92	963.87	8.95E-05
44	0.0251					10.61	51114	0.36							
45		0.0021	0.1019	637915.52	4210.24				113.69	653.08	2318.23	-1051.25	3369.48	958.88	9.53E-05
46	0.0273					10.77	51106	0.38							
47		0.0009	0.0997	638743.77	4215.71				117.80	690.48	2282.19	-1051.25	3333.44	956.05	9.85E-05

Fig 3.7.4: Worksheet for computation of counter recoil velocity and orifice area; variable buffer orifice with constant retardation

	B	C	D	E	F	G	H	I	J	K	L	M	N	O	P
46	0.0273					10.77	51106	0.38							
47		0.0009	0.0997	638743.77	4215.71				117.80	690.48	2282.19	-1051.25	3333.44	956.05	9.85E-05
48	0.0281					10.82	51103	0.38							
49		0.0001	0.0989	639012.40	4217.48				119.15	702.59	2270.51	-1051.25	3321.76	955.14	9.96E-05
50	0.0282					10.83	51103	0.38							
51		0.0002	0.0988	639084.47	4217.96				119.51	705.83	2267.37	-1051.25	3318.63	954.89	9.98E-05
52	0.0284					10.84	51102	0.39							
53		0.0002	0.0986	639169.06	4218.52				119.94	709.64	2263.70	-1051.25	3314.95	954.60	1.00E-04
54	0.0286					10.85	51102	0.39							
55		0.0002	0.0984	639254.32	4219.08				120.36	713.48	2259.99	-1051.25	3311.24	954.31	1.00E-04
56	0.0287					10.86	51101	0.39							
57		0.0002	0.0983	639340.26	4219.65				120.80	717.35	2256.26	-1051.25	3307.51	954.02	1.01E-04
58	0.0289					10.87	51101	0.39							
59		0.0002	0.0981	639426.87	4220.22				121.23	721.25	2252.49	-1051.25	3303.75	953.72	1.01E-04
60	0.0290					10.88	51100	0.39							
61		0.0002	0.0980	639514.18	4220.79				121.67	725.18	2248.70	-1051.25	3299.96	953.43	1.01E-04
62	0.0292					10.89	51099	0.39							
63		0.0002	0.0978	639602.19	4221.37				122.11	729.14	2244.88	-1051.25	3296.13	953.13	1.02E-04
64	0.0293					10.90	51099	0.39							
65		0.0002	0.0977	639690.92	4221.96				122.56	733.13	2241.03	-1051.25	3292.28	952.82	1.02E-04
66	0.0295					10.91	51098	0.39							
67		0.0002	0.0975	639780.36	4222.55				123.01	737.15	2237.15	-1051.25	3288.40	952.52	1.02E-04
68	0.0297					10.92	51098	0.39							
69		0.0027	0.0973	640571.40	4227.77				127.06	772.68	2202.79	-1051.25	3254.04	949.83	1.05E-04
70	0.0324					11.07	51088	0.41							
71		0.0034	0.0946	642262.53	4238.93				135.64	848.38	2129.67	-1051.25	3180.92	944.11	1.12E-04
72	0.0357					11.23	51076	0.43							
73		0.0040	0.0913	644334.49	4252.61				146.22	940.65	2040.51	-1051.25	3091.76	937.14	1.19E-04
74	0.0398					11.38	51061	0.46							
75		0.0049	0.0872	646844.58	4269.17				159.14	1051.72	1933.07	-1051.25	2984.33	928.76	1.28E-04
76	0.0447					11.53	51043	0.48							
77		0.0062	0.0823	649988.54	4289.92				175.55	1189.77	1799.36	-1051.25	2850.61	918.36	1.40E-04
78	0.0508					11.67	51019	0.52							
79		0.0083	0.0762	654127.03	4317.24				197.59	1369.71	1624.70	-1051.25	2675.95	904.82	1.55E-04
80	0.0591					11.79	50988	0.56							
81		0.0121	0.0679	660045.22	4356.30				230.05	1623.55	1377.46	-1051.25	2428.72	885.78	1.77E-04
82	0.0712					11.87	50942	0.61							
83		0.0218	0.0558	670105.45	4422.70				287.82	2045.93	963.71	-1051.25	2014.96	854.21	2.18E-04
84	0.0930					11.86	50856	0.70							
85		0.0068	0.0340	678801.44	4480.09				339.50	2402.09	613.26	-1051.25	1664.51	827.71	2.60E-04
86	0.0998					11.80	50808.6	0.72							
87		0.0068	0.0272	683010.14	4507.87				366.67	2571.59	444.37	-1051.25	1495.63	815.15	2.84E-04
88	0.1066					11.74	50761.2	0.75							
89		0.0068	0.0204	687265.22	4535.95				394.31	2741.08	275.32	-1051.25	1326.58	802.61	3.11E-04
90	0.1134					11.68	50713.8	0.77							
91		0.0068	0.0136	691567.43	4564.35				422.43	2910.57	106.10	-1051.25	1157.35	790.09	3.43E-04
92	0.1202					11.63	50666.4	0.79							
93		0.0068	0.0068	695917.51	4593.06				451.05	3080.07	-63.30	-1051.25	987.96	777.61	3.82E-04
94	0.1270					11.57	50619	0.82							

Fig 3.7.4 (Continued): Worksheet for computation of counter recoil velocity and orifice area; variable buffer orifice with constant retardation

The computations for the buffing phase complete, it is possible to compute the time duration and plot the travel-velocity curves for the entire recoil cycle.

Duration of Recoil Cycle and Rate of Fire

From recoil computation recoil duration = 0.098s
From counter recoil computation duration of counter recoil: 1.31s
From buffing computation, duration of buffing for Case 1 = 0.248s
From buffing computation, duration of buffing for Case 2 = 0.311s

Total recoil cycle duration for Case 1 of buffing = 1.656s
Total recoil cycle duration for Case 2 of buffing = 1.719s

Hence the rate of fire achievable from recoil duration consideration alone is about 35 rounds per minute.

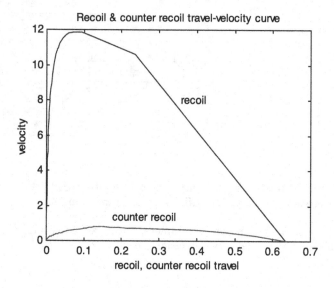

Fig: 3.7.5 Travel-velocity relation for the entire recoil cycle

3.8 Design of Recoil System Components

General Requirements

Recoil system components are subject to stresses of considerable magnitude and varying direction over small intervals of time. Usually recoil systems are a mix of hydraulic and pneumatic sub systems wherein sealing is critical to the efficient functioning of the system under adverse conditions of service. Recoil systems also contribute to the overall weight of the weapon hence the components cannot be overly heavy. Hence strength, resistance to dilation and lightness are important requirements of recoil system components.

Design of Recoil System Piston Rods

Recoil piston rods are tension members and their design depends on the yield strength of the material used as also the maximum force on the pistons to which attached. The maximum force on the piston being known, the piston rod area is obtained by dividing the maximum force on the piston by the yield strength of the material:

$$A_{Br} = \frac{F_{B0}}{\sigma_{yp}} \quad\text{.. [3.8.1]}$$

The calculation is illustrated with the help of the following example:

Example 3.8.1

Calculate the area of the piston rods of the recoil brake and recuperator cylinder given the following information:

Maximum hydraulic force on the recoil brake piston: 50,000.00 N
Initial recuperator force: 5000 N
Final recuperator force is twice initial recuperator force.
Yield strength of the steel used: 300.0 M Pa

Minimum area of the recoil brake piston rod

$$A_{Br} = \frac{F_{B0}}{\sigma_{yp}} = \frac{50000.00}{300.0.10^6} = 1.67.10^{-4}\,\text{m}^2$$

Hence diameter of the piston rod:

$$d_{Br}^2 = \frac{1.67.10^{-4}.4.0}{\pi}$$

Or:

$$d_{Br} = 14.56\,\text{mm}$$

Minimum area of the recuperator piston rod

Maximum force on the recuperator piston:

$$F_{Rf} = 2.5000.00 \;\; \text{N}$$

$$A_{Rr} = \frac{10000.00}{300.10^6} = 3.33.10^{-5}\,\text{m}^2$$

Diameter of the recuperator piston rod is:

$$d_{Rr}^2 = \frac{3.33.10^{-5}.4}{\pi}$$

$$d_{Rr} = 6.51\,\text{mm}$$

Design of Recoil Brake Piston

The minimum area of the recoil brake piston is dictated by the pressure capacity of the seal. The area of the recoil brake piston is given by:

$$A_{Bp} = \frac{F_{B0}}{p_s} \quad \text{..} \quad [3.8.2]$$

p_s: pressure capacity of the seal

The diameter of the recoil brake piston is now:

$$d_{Bp} = \sqrt{\frac{4A_{Bp}}{\pi} + d_{Br}^2} \quad \text{..} \quad [3.8.3]$$

Example 3.8.2

Determine the minimum diameters of the recoil brake pistons given the data in Example 3.8.1 above given that the pressure capacity of the seals is 0.03 G Pa.

Solution

Recoil brake piston diameter

Recoil brake piston area is:

$$A_{Bp} = \frac{50000.00}{0.03.10^9} = 1.6.10^{-3}\,\text{m}^2$$

Piston diameter is:

$$d_{Bp} = \sqrt{\frac{1.6.10^{-3}.4}{\pi} + \left(14.57.10^{-3}\right)^2} = 48.3\,\text{mm}$$

Design of Cylinders

The design of the recoil brake cylinder follows the method for design of thick cylindrical high pressure vessels. The inner diameter of the cylinders is defined by piston dimensions established by Equation [3.8.3]. The thickness of the cylinder depends upon the pressure capacity of the seals and the yield strength of the material.

215

For design purposes a higher pressure called the proof pressure is used which is about 1.5 times the pressure capacity of the seal. Hence:

$$p_p = 1.5 p_s \quad \text{..} \quad [3.8.4.]$$

From the Maximum Shear Stress Theory:

$$\sigma_t - \sigma_r = \sigma_{yp}$$

σ_t : hoop stress

σ_r : radial stress

σ_{yp} : yield stress of the material

A safety factor of around 1.5 is usually introduced so that the material is stressed below its yield point under the proof stress. The equation immediately above becomes:

$$\sigma_t - \sigma_r = \frac{\sigma_{yp}}{SF} \quad \text{..} \quad [3.8.5]$$

SF: Safety Factor

Because $\sigma_r = -p_p$, Equation [3.8.5] yields:

$$\sigma_t = \frac{\sigma_{yp}}{SF} - p_p \quad \text{..} \quad [3.8.6]$$

From the analysis for stress in a thick walled cylinder:

$$\sigma_t = p_p \left(\frac{T_c^2 + 1}{T_c^2 - 1} \right)$$

D: outer diameter of the cylinder

d: inner diameter of the cylinder

$T_c = \dfrac{D}{d}$: wall ratio of the cylinder

216

From which:

$$T_c = \sqrt{\frac{\dfrac{\sigma_t}{p_p} + 1}{\dfrac{\sigma_t}{p_p} - 1}} \quad \text{..} \quad [3.8.7]$$

The calculations are demonstrated with the help of the following example.

Example 3.8.3

Calculate the outer diameter for the recoil brake cylinder using the data and results of Examples 3.8.1 and 3.8.2.

Solution

Proof pressure for the recoil brake cylinder as given by Equation [3.8.4]:

$$p_p = 1.5 p_s = 1.5.0.03.10^9 = 0.045 \text{ G Pa}$$

Hoop stress:

$$\sigma_t = \frac{300.10^6}{1.5} - 0.045.10^9 = 155.0 \text{ M Pa}$$

Wall ratio of the recoil brake cylinder is:

$$T_B = \sqrt{\frac{\dfrac{155}{45} + 1}{\dfrac{155}{45} - 1}} = 1.82$$

From which outer diameter of the recoil brake cylinder is:

$$d_{Bco} = 1.82.48.3 = 87.82 \text{ mm}$$

Design of Recuperator Cylinder

Selection of Initial Pressure

The area of the recuperator cylinder is determined from the initial recuperator force and the initial pressure in the recuperator. The initial pressure is fixed based on the pressure of available source of gas supply, which is usually used in the filling of high-pressure cylinders. For efficient filling into the recuperator the initial pressure difference should be high. An initial pressure of around half the pressure in the supply source may be selected. Once the initial pressure is selected, the area of the recuperator piston is calculated such that the initial recuperator force requirement to keep the recoiling parts in the firing position at the maximum angle of elevation is met. The initial recuperator force is given by Equation [3.4.3] Reference 1.

The area of the recuperator piston is now determined from:

$$A_{R_p} = \frac{F_{r(0)}}{p_{r(0)}} \quad \text{.. [3.8.8]}$$

$F_{r(0)}$: initial recuperator force
$p_{r(0)}$: initial recuperator pressure

The final pressure is selected, as mentioned earlier, around 1.5 to 2.0 times the initial pressure, depending on the desired degree of energetic return, to the firing position, of the recoiling parts. The following relations having been established, it is possible to proceed with the design of the recuperator cylinder.

$$p_{rf} = \alpha p_{r(0)} \; ; \quad \alpha = 1.5 \text{ to } 2.0$$

$$V_0 = V_f + \delta V \; ; \text{ Where } \delta V = A_{R_p} l_r$$

p_{Rf}: recuperator final pressure
V_0: initial volume of gas in recuperator
V_f: final volume of gas in recuperator
A_{Rp}: recuperator piston area
l_r: recoil length

From the gas laws:

$$\frac{p_{Rf}}{p_{R(o)}} = \left(\frac{V_0}{V_0 - \delta V}\right)^n$$

n: gas exponent

From which:

$$V_0 = \frac{\alpha^{\frac{1}{n}} \delta V}{\alpha^{\frac{1}{n}} - 1} \quad \dots\dots\dots\dots\dots\dots\dots\dots\dots\dots\dots\dots\dots\dots\dots\dots\dots\dots\dots\dots\dots\dots\dots\dots\dots \quad [3.8.9]$$

Example 3.8.4

Determine the area of the recuperator piston given that the initial recuperator pressure is 0.6 M Pa and the initial recuperator force is 4116.75 N. The final recuperator pressure is twice the initial pressure, value of the gas exponent is 1.3 and the diameter of the piston rod is $9.2.10^{-3}$ m. Also find the thickness of the cylinder wall. The pressure capacity of the recuperator seal is 27.5 M Pa. The recoil length is 0.635 m. Yield strength of steel selected is 300.0 M Pa.

Solution

The area of the recuperator piston is given by Equation [3.8.8]:

$$A_{Rp} = \frac{4116.75}{0.6.10^6} = 6.86.10^{-3} \, \text{m}^2$$

Vide Equation [3.8.9], the initial volume of gas in the recuperator cylinder is:

$$V_0 = \frac{2^{\frac{1}{1.3}}.6.86.10^{-3}.0.635}{2^{\frac{1}{1.3}} - 1} = 0.01054 \, \text{m}^3$$

The final volume of gas in the recuperator cylinder is:

$V_f = 0.01054 - 6.86.10^{-3}.0.635 = 6.18.10^{-3} \, \text{m}^3$

Thickness of the cylinder wall

Proof pressure for the recuperator cylinder as given by Equation [3.8.4]:

$p_p = 1.5 p_s = 1.5.27.5.10^6 = 41.25 \, \text{M Pa}$

Hoop stress from Equation [3.8.6]:

$\sigma_t = \dfrac{300.10^6}{1.5} - 41.25.10^6 = 158.75 \, \text{M Pa}$

Wall ratio of the recuperator cylinder is:

$$T_B = \sqrt{\dfrac{\dfrac{158.75}{41.25}+1}{\dfrac{158.75}{41.25}-1}} = 1.3$$

Applying Equation [3.8.3], the diameter of the recuperator piston rod works out to be 93.7 mm.

From which outer diameter of the recuperator cylinder is:

$d_{Rco} = 1.3.0.0937 = 121.79 \, \text{mm}$

4

Design of Balancing Gears

4.1 Design Considerations of Balancing Gears

Requirement for Balancing Gears

The need for balancing has been dealt with in Reference 1, wherein it was established that the most desirable method of balancing is by introducing a balancing couple which varies in order to equal the out of balance moment inherent in equipments with rear trunnions.

Selection of a Balancing Gear

The selection of a balancing gear cannot be done in isolation. Space and the geometrical layout of the weapon are the primary technical factors leading to the choice of type and configuration of a balancing gear. The balancing gear has to be fitted into the available space, with adjustments where necessary. The spring deflection or the piston rod displacement, in the case of pneumatic balancing gears, has to be within the constraints of space of the weapon. The volume and weight of the balancing gear may also be inhibited by the same constraints. Other factors, which lead to an overall successful design, are simplicity of manufacture and maintenance, reliability and minimum cost.

Factors Affecting Design of a Balancing Gear

The selection of the balancing gear having been made, the factors which dictate its capacity and final configuration, are as follows:

Out of Balance Moment: The magnitude of the out of balance moment demands an equal and opposite moment called the balancing moment to negate it. The out of balance moment is given by:

$$M_{ob} = W_e a Cos\phi \quad .... [4.1.1]$$

W_e: weight of elevating parts
a: moment arm of weight of elevating parts about trunnion axis
φ : angle of elevation

More often than not, the centre of gravity of the elevating parts will not lie on the axis of the bore. In such an eventuality, the actual angle needs to be considered when calculating the out of balance moment.

It is possible in theory, to obtain perfect balance by varying the balancing moment, or its moment arm, so that their product varies in accordance with $Cos\varphi$, provided certain initial conditions are met. Practically, however, perfect balance is almost impossible to achieve and the designer has to be content with the closest approximation possible.

Friction: The criterion for perfect balance was arrived at in Reference 1 without inclusion of the effect of friction on the balancing. In a balancing gear, the effect of friction cannot be ignored. Further, it cannot be catered for by the balancing gear itself, due to the fact that the friction force changes direction as the elevating parts are elevated or depressed. Friction in a balancing gear arises at the trunnions, in the linkages at the cradle and in the linkages at the bottom carriage. It also occurs in the pneumatic and hydraulic cylinders as a result of the friction of the seals. Friction is also present between the inner and outer cylinders of balancing gears, which are based on springs in compression.

Geometry and Positioning of the Balancing Gear: The positioning of the balancing gear with reference to the trunnions is critical towards achieving acceptable balancing. Hence the position, which meets the requirement of perfect balance, is first established. This may be modified subsequently but by remaining within the limits of acceptability as described in the next paragraph.

Acceptable Degree of Balancing

As perfect balance is not often realizable, the degree of imbalance should be small enough to be overcome by reasonable effort at the hand wheel. Weapons even if possessing power elevation are designed to be amenable to manual elevation to cater for contingencies of non availability of power due to tactical or technical reasons.

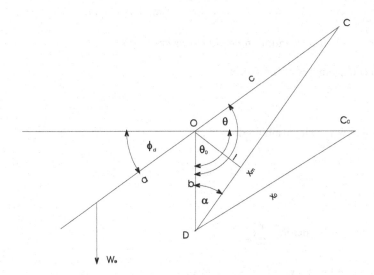

Fig 4.1.1: Balancing gear geometry at position of maximum angle of depression

Establishing the Minimum Energy Capacity of a Balancing Gear

It is always desirable to have a compact balancing gear. To meet his requirement it must be ensured that the maximum energy stored in the spring of the balancing gear is the minimum necessary for it to function effectively. This is possible by careful choice of the balancing gear geometry.

Fig 4.1.1 depicts a balancing gear of constant spring rate. For simplicity, the centre of gravity of the elevating parts is assumed to lie on the axis of the bore. When this is not the case, the actual angle may be applied when calculating the out of balance moment.

The maximum energy is stored in the spring at its maximum deflection, which is at the maximum angle of depression of the elevating parts. From the geometry of the diagram and applying the Cosine rule:

$$x_m^2 = b^2 + c^2 - 2bcCos\theta \qquad\qquad\qquad [4.1.2]$$

x_m: spring length at maximum angle of depression

Also from the geometry of the figure:

$$\theta = \theta_0 + \phi_d$$

θ_0 : value θ of when elevation angle is zero

Hence:

$$Cos\theta = Cos(\theta_0 + \phi_d)$$

Since for perfect balance $\theta_0 = \dfrac{\pi}{2}$:

$$Cos\theta = Cos(\theta_0 + \phi_d) = -Sin\phi_d$$

Equation [4.1.2] can be rewritten as:

$$x_m^2 = b^2 + c^2 + 2bcSin\phi_d \dots\dots\dots\dots\dots\dots\dots\dots [4.1.3]$$

φ_d : angle of maximum depression

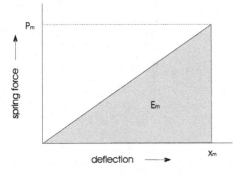

Fig 4.1.2: Spring force-deflection relation

From the spring force-deflection graph depicted in Fig 4.1.2 above, the maximum energy stored in the system is:

$$E_m = \frac{1}{2}P_m x_m$$

P_m: spring force at maximum deflection
x_m: maximum spring deflection

If P_0 is the initial spring force at initial deflection x_0 and s its stiffness factor, then:

$$P_m = P_0 + s(x_m - x_0)$$

Also:

$$P_0 = sx_0$$

Hence:

$$P_m = sx_m$$

And:

$$E_m = \frac{1}{2}sx_m^2$$

Again for perfect balance, the spring stiffness factor is: $s = \dfrac{W_e a}{bc}$

Substituting for s and x_m:

$$E_m = \frac{1}{2}\frac{W_e a}{bc}\left(b^2 + c^2 + 2bcSin\phi_d\right)$$

Or:

$$E_m = \frac{W_e a}{2}\left(\frac{b}{c} + \frac{c}{b} + 2Sin\phi_d\right)$$

Substituting for $\dfrac{b}{c} = K$

$$E_m = \frac{W_e a}{2}\left(K + \frac{1}{K} + 2Sin\phi_d\right) \quad\text{... [4.1.4]}$$

In the above equation the variable E_m is expressed in terms of K, also a variable hence the minimum or maximum value of E_m may be established by the rules of maxima and minima.

Differentiating Equation [4.1.4] once:

$$\frac{dE_m}{dK} = \frac{W_e a}{2}\left(1 - \frac{1}{K^2}\right)$$

Equating the right hand side of the equation to zero:

$$\frac{W_e a}{2}\left(1-\frac{1}{K^2}\right)=0$$

The roots of the equation are:

$$K=\pm 1$$

In this situation, negative values of K are insignificant as neither b nor c can assume negative values, so the root $K=1$ is accepted.

Differentiating the equation in question for the second time:

$$\frac{d^2 E_m}{dK^2}=\frac{W_e a}{2}\frac{1}{K^3}$$

Since $\frac{d^2 E_m}{dK^2}$ is positive for $K=1$, the value of E_m when $K=1$ is minimum.

Relation between b/c ratio and Maximum Energy of a Balancing Gear

The relation between the ratio of $\frac{b}{c}$ and the maximum energy stored in a balancing gear can be seen from the following example:

Example 4.1.1

The muzzle preponderance of a gun is 2900 N. The moment arm of the out of balance moment is 0.5 m. Plot the relation between $\frac{b}{c}$ ratios from 0.25 to 10 and verify the value of $\frac{b}{c}$ for which the energy in the balancing gear spring is the minimum. The angle of maximum depression of the gun is 5°.

Solution

The problem is conveniently solved with the help of the Matlab programme given below.

Matlab Programme 4.1.1

```
% programme to compute Em-K relation
% file name: Em_K.txt

We=200*9.91 % muzzle preponderance N
a=.5 % moment arm of We m
K=linspace(.1,10,100)
Em=We.*a./2*(K+1./K.^2+2*sin(5./180.*pi))
plot(K,Em)
title('b/c-Maximum energy in balancing gear relation')
xlabel('b/c');ylabel('Em')
```

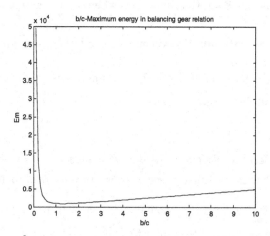

Fig 4.1.3: $\dfrac{b}{c}$ ratio –maximum energy relation of a balancing gear

228

From Fig 4.1.3, the least magnitude of the maximum stored energy in a spring balancing gear is at a $\frac{b}{c}$ ratio of 1. However, the increase in the stored energy is not significant for $\frac{b}{c}$ ratios less than around 3.

Imperfectly Balanced Systems

When physical constraints do not permit the balancing gear geometry necessary for attainment of perfect balance, the balancing gear is designed to approximate perfect balance over the entire elevation range of the weapon.

The theoretical balancing force is obtained by dividing the out of balance moment about the trunnion axis, at intervals of elevation angle over the elevation range of the weapon, by the balancing moment arm about the same axis. The theoretical balancing force so determined will not fall on a straight line. Hence it is necessary to determine the equation of a straight line, which best represents, the aggregate of the data in question. Two methods by which this is achieved are covered subsequently.

Approximate Method for Determining Spring Stiffness and Initial Spring Force of a Spring Type Balancing Gear

From the geometry of Fig 4.1.1, the length of the balancing gear spring at an angle of elevation φ is given by:

$$x^2 = b^2 + c^2 - 2bcCos\theta$$

But $\theta = \frac{\pi}{2} - \phi$

Hence:

$$x^2 = b^2 + c^2 - 2bcSin\phi \quad .... [4.1.5]$$

Also from Fig 4.1.1:

$$\frac{c}{Sin\alpha} = \frac{x}{Sin\theta}$$

From which:

$$Sin\alpha = \frac{cSin\theta}{x}$$

But $Sin\alpha = \frac{l}{b}$

l: moment arm of the balancing gear force

Finally:

$$l = \frac{bcCos\phi}{x} \quad \dots\dots\dots\dots\dots\dots\dots\dots\dots\dots\dots\dots\dots\dots\dots\dots\dots\dots\dots\dots\dots\dots\dots\dots\dots\dots\dots\dots\dots [4.1.6]$$

In order to determine the initial balancing gear force, also called the assembled spring force and the spring stiffness factor, the following method is employed:

The initial force of the balancing gear taken at an elevation angle of zero is given by:

$$P_0 = \frac{W_e a}{l_0}$$

l_0: moment arm of the balancing gear force at zero elevation, calculated from Equation [4.1.6]

At the maximum angle of elevation of the equipment ϕ_m, the balancing gear force is:

$$P_m = \frac{W_e aCos\phi_m}{l_m}$$

l_m: moment arm of the balancing gear force at the maximum angle of elevation ϕ_m, calculated from Equation [4.1.6].

The Spring Stiffness Factor is now:

$$s = \frac{P_0 - P_m}{x_0 - x_m} \quad \text{...} \quad [4.1.7]$$

The above method is illustrated in the following worked example.

Example 4.1.2

The following data applies to a gun for which a spring type balancing gear is to be designed. Muzzle preponderance is 2900 N. With reference to Fig 4.1.1; the lengths b and c are both equal to 1.0 m. The moment arm a is 0.5 m. The gun has an elevation range of 0° to +70°. Establish approximate values for the initial balancing force and the Spring Stiffness Factor.

Spring length at zero elevation:

$$x_0 = \sqrt{b^2 + c^2} = \sqrt{1+1} = 1.414 \text{ m}$$

Moment arm of the balancing force at zero elevation:

$$l_0 = \frac{bc}{x_0} = \frac{1}{1.414} = 0.707 \text{ m}$$

Initial balancing force

$$P_0 = \frac{W_e a}{l_0} = \frac{2900.0.5}{0.707} = 2050.92 \text{ N}$$

Moment arm of balancing force at 60° elevation

Spring length at 60° elevation:

$$x_{60}^2 = b^2 + c^2 - 2bcSin60 = 1 + 1 - 2x0.866 = 0.268 \text{ m}$$

$$x_{60} = 0.518 \text{ m}$$

$$l_{60} = \frac{bcCos60}{x_{60}} = \frac{0.5}{0.518} = 0.966 \text{ m}$$

Balancing force at 60° elevation

$$P_{60} = \frac{W_e aCos60}{l_{60}} = \frac{2900.0.5.0.5}{0.966} = 750.51 \text{ N}$$

Spring Stiffness Factor

$$s = \frac{P_0 - P_{60}}{x_0 - x_{60}} = \frac{2050.92 - 750.51}{1.414 - 0.518} = 1451.342 \text{ N/m}$$

Determination of Initial Balancing Force and Spring Stiffness Factor by Method of Least Squares

The balancing force of the balancing gear is given by:

$$P = P_0 - s(x_0 - x)$$

Or: $P = P_0 - s\delta x$

The error in P for every set of data $\delta x_i, P_i$ can be defined as:

$$e_i = P_i - f(\delta x_i)$$

Or:

$$e_i = P_i - (s\delta x_i + P_0)$$

Hence the sum of the squares of the errors is given by:

$z = \sum_{i=1}^{i=n} e_i^2$, where there are n data points.

Or:

$$z = \sum_{i=1}^{n}\left[P_i - \left(s\,\delta x_i + P_0\right)^2\right]$$

In order to determine the values of s and P_0, which will minimize z, the derivatives of z with respect to s and P_0 must be equated to zero. Hence:

$$\frac{\partial z}{\partial P_0} = -2\sum_{i=1}^{n}\left[P_i - \left(s\,\delta x_i + P_0\right)\right] = 0$$

$$\frac{\partial z}{\partial s} = -2\sum_{i=1}^{n}\delta x_i\left[P_i - \left(s\,\delta x_i + P_0\right)\right] = 0$$

It follows that:

$$\sum_{i=1}^{n}\left(s\,\delta x_i + P_0 - P_i\right) = 0$$

And:

$$\sum_{i=1}^{n}\left(s\,\delta x_i^2 + P_0\,\delta x_i - P_i\,\delta x_i\right) = 0$$

Finally:

$$s\sum_{i=1}^{n}\delta x_i + nP_0 = \sum_{i=1}^{n}P_i \quad\text{...} \quad [4.1.8]$$

$$s\sum_{i=1}^{n}\delta x_i^2 + P_0\sum_{i=1}^{n}\delta x_i = \sum_{i=1}^{n}P_i\,\delta x_i \quad\text{...................................} \quad [4.1.9]$$

Solution of the above simultaneous equations yields the spring stiffness and the initial balancing force. The methodology is illustrated in the following example:

Example 4.1.3

Given the data contained in Example 4.4.2, compute by the method of least squares the equation of the most appropriate balancing force line.

Solution

The theoretical balancing force is calculated at intervals of $10°$ elevation with the help of Equations [4.1.6] and by equating the out of balance and balancing moments.

The spring displacement is calculated with the help of Equation [4.1.5] and the change in length δx calculated.

Applying Equations [4.1.8] and [4.1.9], two equations in spring stiffness s and initial spring force P_0 are generated:

$4.07s + 8.00P_0 = 13504.36$
$3.06s + 4.07P_0 = 6363.54$

These may be solved for s and P_0 by two methods using inbuilt Excel features. The first is by the method of matrix inversion. The second is by use of the Trendline feature. Both methods are illustrated in an Excel worksheet.

Trendline Feature in Excel

A linear trendline is a best-fit straight line that is used with simple linear data sets. The data is linear if the pattern in its data points resembles a line. A linear trendline is a best-fit straight line that is used with simple linear data sets. Your data is linear if the pattern in its data points resembles a line. A linear trendline shows that something is increasing or decreasing at a steady rate.

The use of the Trendline feature in Excel to solve the problem is also illustrated. The method is outlined ahead:

Highlight the data series i.e. P and dx holding down the control key, which the trendline is to be added.
Click on the chart menu.
Choose XY (scatter) from the standard types.
Choose scatter as the chart sub type. Click finish. The scatter chart is generated.
Click on any of the data points. Click on the add trendline option.
On the type tab, click the linear type of regression trendline.
On the options tab check the display equation on chart box

	A	B	C	D	E	F	G	H
1		Spreadsheet to determine initial balancing gear force, spring stiffness factor						
2			b m	1	We N	2900.00	n	8
3			c m	1	a m	0.50		
4	phi	phi(rad)	x	dx	l	P	dx^2	P.dx
5	0	0.00	1.41	0.00	0.71	2050.61	0.00	0.00
6	10	0.17	1.29	0.13	0.77	1864.08	0.02	239.79
7	20	0.35	1.15	0.27	0.82	1663.37	0.07	444.22
8	30	0.52	1.00	0.41	0.87	1450.00	0.17	600.61
9	40	0.70	0.85	0.57	0.91	1225.59	0.32	697.33
10	50	0.87	0.68	0.73	0.94	991.86	0.53	724.23
11	60	1.05	0.52	0.90	0.97	750.58	0.80	672.95
12	70	1.22	0.35	1.07	0.98	503.58	1.14	537.28
13		sigma		4.07		10499.67	3.06	3916.41
14		A		A inv		B	X	AX
15		4.07	8.00	-0.51668	1.01494	10499.67	-1450.00	10499.67
16		3.06	4.07	0.38802	-0.51668	3916.41	2050.61	3916.41

Fig 4.1.4: Worksheet to determine initial spring force and spring stiffness factor

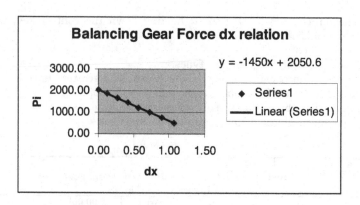

Fig 4.1.5: Use of the Trendline feature in Excel for determination of Spring Stiffness Factor and assembled spring force

4.2 Pneumatic Balancing Gears

Balancing Moment of a Frictionless Pneumatic Balancing Gear

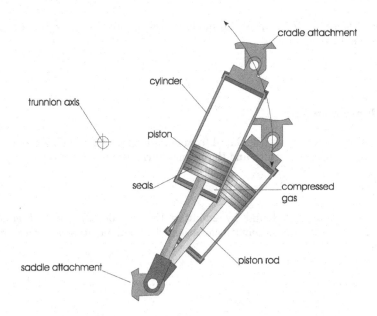

Fig 4.2.1: Sectionized view of a typical pneumatic balancing gear

Calculation of the Balancing Gear Moment Arm

The first step in balancing gear design is the calculation of the balancing moment arm. With reference to Fig 4.1.1, when the angle of elevation is φ and b and c are known, vide Equation [4.1.3], for angles of depression, the length of the spring is:

$$x^2 = b^2 + c^2 + 2bcSin\phi$$

Vide Equation [4.1.5], for all angles of elevation:

$$x^2 = b^2 + c^2 - 2bcSin\phi$$

The length of the moment arm of the balancing gear force can now be determined from Equation [4.1.6]:

$$l = \frac{bc}{x}Cos\phi$$

Out of Balance Moment

The out of balance moment is calculated from equation [4.1.1], knowing the elevation range, the muzzle preponderance and the moment arm of the balancing force.

$$M_{ob} = W_e a Cos\phi$$

The initial gas volume is found by equating the out of balance and balancing moments at zero and at maximum elevation. The balancing gear force at zero elevation is:

$$P_0 = \frac{W_e a}{l_0} \dotfill [4.2.1]$$

And the pressure is:

$$p_0 = \frac{P_0}{A}$$

A: balancing gear piston area

The absolute pressure is:

$$p_{a0} = p_0 + p_{atm} \dotfill [4.2.2]$$

p_{atm}: atmospheric pressure

238

The balancing gear force at maximum elevation is:

$$P_m = \frac{W_e a}{l_m}$$

The pressure in this case is:

$$p_m = \frac{P_m}{A} \quad \text{...} \quad [4.2.3]$$

The absolute pressure being:

$$p_{m0} = p_m + p_{atm} \quad \text{...} \quad [4.2.4]$$

Determination of Initial Gas Volume

The initial gas volume, when the gas process during elevation or depression is isothermal, as in manual elevation and depression, is determined by applying Boyle's law:

$$p_0 V_0 = p_m V_m$$

$V_m = V_0 + \delta V_m$: gas volume at maximum elevation

Hence:

$$V_0 = \frac{\dfrac{p_m}{p_0} \delta V_m}{1 - \dfrac{p_m}{p_0}} \quad \text{...} \quad [4.2.5]$$

Calculation of the Balancing Moment for a Frictionless Pneumatic Balancing Gear

The change in gas volume at any angle of elevation is given by:

$\delta V = A \delta x$

Hence the gas volume, at the angle of elevation under consideration, is:

$V = V_0 + \delta V$

The absolute pressure is:

$p_a = \dfrac{p_{a0} V_0}{V}$ and the actual pressure: $p = p_a - p_{atm}$

The balancing gear force is hence:

$P = pA$

The balancing moment is given by:

$M_b = Pl$ .... [4.2.6]

Fig 4.2.1 depicts the out of balance and balancing moments for a typical pneumatic type balancing gear without the effect of friction force. The difference between the two moments is acceptable if it does not exceed the manual torque, which can be applied at the hand wheel without undue effort.

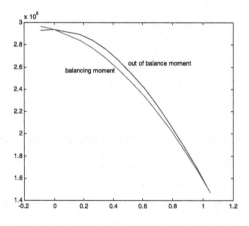

Fig 4.2.2: Out of balance and balancing moment against angle of elevation for a typical pneumatic balancing gear

Friction Calculations

Friction forces that arise during balancing gear operation are due to force exerted by the seals of the piston and the piston rod on the balancing gear cylinder and friction at the bearing at the points of attachment to the bottom carriage and to the cradle. In the case of pneumatic balancing gears both friction forces have to estimated and reflected in the final design calculations.

Seal Friction Force

The friction force due to seal pressure of the piston on the cylinder walls and around the piston rod is computed as in the case of seals for other hydro pneumatic systems such as the recoil system as follows:

The spring pressure of the seals is given by:

$$p_{sp} = \frac{\upsilon - K_p}{K_p} p_m$$

υ : Leakage Factor
K_p: Pressure Factor
p_m: maximum gas pressure in the balancing gear cylinder

The friction force due to the piston seal is given by:

$$F_p = \mu \pi d_p w K_p \left(p + p_{sp} \right) \quad\quad\quad [4.2.7]$$

μ : coefficient of friction of seal on wall
d_p: piston diameter
w_p: piston seal width
p: gas pressure at angle of elevation under consideration

The friction force due to the piston rod seal is given by:

241

$$F_r = \mu \pi d_r w K_p \left(p + p_{sp} \right) \quad \text{..} \quad [4.2.8]$$

μ : coefficient of friction of seal on wall
d_r: piston rod diameter
w_r: piston rod seal width
p: gas pressure at angle of elevation under consideration

The net friction seal force is now:

$$F_s = F_p + F_r \quad \text{...} \quad [4.2.9]$$

Net Balancing Gear Force

The effective balancing gear force after allowing for seal friction force is:

$$P_e = P - F_s \quad \text{...} \quad [4.2.10]$$

Bearing Friction at the Points of Attachment during Elevation

The friction torque at the point of attachment to the bottom carriage is:

$$T_D = \mu P_e r_D \quad \text{...} \quad [4.2.11]$$

μ : coefficient of friction
r_D: radius of bottom carriage bearing

The friction torque at the point of attachment to the cradle is:

$$T_C = \mu P_e r_C \quad \text{...} \quad [4.2.12]$$

μ : coefficient of friction
r_c: radius of bottom carriage bearing

The friction force to be overcome by the balancing gear due to the friction at the attachment bearings of the bottom carriage and cradle is computed considering the diagram of the balancing gear isolated as below separately during elevation and depression:

Friction Force at Attachment Bearings during Elevation

In Fig 4.2.3 below, D is the point of attachment of the balancing gear with the saddle and C with the cradle. During elevation the balancing gear moves clockwise with respect to the pivot at D and clockwise with respect to the pivot at C. The friction torques are hence both anticlockwise.

Friction Force Due to Friction Torque at Bottom Carriage Bearing

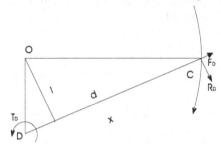

Fig 4.2.3: Friction force in the balancing gear during elevation due to friction torque at bottom carriage pivot

R_D: reaction at C due to the friction torque at D
F_D: friction force in balancing gear due to bearing friction at D

Taking moments about D:

$$R_D x = T_D, \text{ or: } R_D = \frac{T_D}{x}$$

Taking moments about the trunnion axis O:

$$R_D d = F_D l + T_D$$

From which the friction force, due to friction at the bottom carriage bearing, in the balancing gear during elevation is:

$$F_{D_e} = \frac{T_D}{l}\left(\frac{d}{x} - 1\right)$$ .... [4.2.13]

The subscript e here denotes elevation.

Friction Force Due to Friction Torque at Cradle Bearing

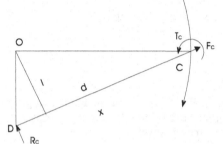

Fig 4.2.4: Friction force in the balancing gear during elevation due to friction torque at cradle bearing

R_C: reaction at D due to the friction torque at C
F_C: friction force in balancing gear due to bearing friction at C

Taking moments about C:

$$R_C x = T_C, \text{ or: } R_C = \frac{T_C}{x}$$

Taking moments about the trunnion axis O:

$$F_c l + T_C = R_C(x - d)$$

Or:

$$F_{C_e} = -T_C \frac{d}{lx}$$ .... [4.2.14]

244

The net friction force is:

$$F_{f_e} = F_{D_e} + F_{C_e} \quad \ldots\ldots\ldots\ldots\ldots\ldots\ldots\ldots\ldots\ldots\ldots\ldots\ldots\ldots\ldots\ldots\ldots\ldots\ldots\ldots\ldots\ldots \text{[4.2.15]}$$

Friction Force at Attachment Bearings during Depression

During depression the balancing gear moves anti-clockwise with respect to the pivot at D and anti-clockwise with respect to the pivot at C. The friction torques are hence both clockwise.

Friction Force Due to Friction Torque at Bottom Carriage Bearing during Depression

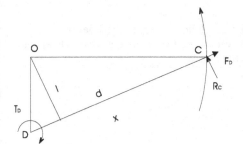

Fig 4.2.5: Friction force in the balancing gear during depression due to friction torque at bottom carriage bearing

R_D: reaction at C due to the friction torque at D
F_D: friction force in balancing gear due to bearing friction at D

Taking moments about D:

$$R_C x = T_D \text{ , or: } R_C = \frac{T_D}{x}$$

Taking moments about the trunnion axis O:

$$F_D l + R_C d = T_D$$

245

From which the friction force due to friction at the bottom carriage attachment, in the balancing gear during depression is:

$$F_{D_d} = \frac{T_D}{l}\left(1 - \frac{d}{x}\right)$$ .... [4.2.16]

The subscript d here denotes depression.

Friction Force Due to Friction Torque at Cradle Bearing during Depression

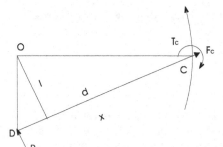

Fig 4.2.6: Friction force in the balancing gear due to friction torque at cradle bearing during depression

R_C: reaction at D due to the friction torque at C
F_C: friction force in balancing gear due to bearing friction at C

Taking moments about C:

$$R_C x = T_C \text{ , or: } R_C = \frac{T_C}{x}$$

Taking moments about the trunnion axis O:

$$T_C = F_C l + R_C(x - d)$$

Or:

$$F_{C_d} = T_C \frac{d}{lx} \quad \text{..} \quad [4.2.17]$$

The net friction force is now:

$$F_{f_d} = F_{D_d} + F_{C_d} \quad \text{..} \quad [4.2.18]$$

Manual Elevation and Depression

The balancing force is in the case of manual elevation and depression given by:

$$P_b = P_e - F_f \quad \text{...} \quad [4.2.19]$$

The term F_f depends, without further elaboration, on whether the movement is in elevation, Equation [4.2.15], or depression, Equation [4.2.18].

The balancing moment is given by:

$$M_{b_{man}} = P_b l \quad \text{..} \quad [4.2.20]$$

The torque required for elevatation is:

$$T_{e_{man}} = M_{ob} - M_{b_{man}} \quad \text{...} \quad [4.2.21]$$

The torque required for depression is:

$$T_{d_{man}} = M_{ob} - M_{b_{man}} \quad \text{...} \quad [4.3.21a]$$

The sub-sub script *man* here implies manual elevation or depression.

Power Elevation and Depression

Behaviour of the Gas

During power elevation and depression, the expansion or compression of the gas is polytrophic unlike in the case of manual elevation and depression where it is isothermal. The calculations are identical to those in the case of manual elevation and depression except for the law applied for determining the pressures at different angles of elevation.

Balancing Moment during Power Elevation

In the case of polytrophic expansion during power elevation, the polytrophic gas law applied is:

$$p_a V^n = p_{a0} V_0^n$$

From which the absolute gas pressure p of volume V of the gas at the angle of elevation under consideration, is given by:

$$p_a = p_{a0} \left(\frac{V_0}{V} \right)^n \quad \text{...[4.2.22]}$$

The actual gas pressure now is:

$$p = p_a - 101300 \text{ Pa} \quad \text{...[4.2.23]}$$

The gas force on the piston is:

$$P = pA$$

The effective force available for balancing the out of balance moment is:

$$P_e = P - F_s \quad \text{..[4.2.24]}$$

F_s: seal friction force

The friction force due to friction at the bearing is calculated similarly as in the case of manual elevation and applied to the force in the equation immediately above to determine the final balancing moment.

Balancing Moment during Power Depression

Assuming that the period between power elevation and the beginning of power depression is adiabatic, the pressure volume relationship during power depression will be identical to that during power elevation. The absolute gas pressure during depression is computed applying the following gas law:

$$p_a V^n = p_{am} V_m^n$$

Or:

$$p_a = p_{a_m} \left(\frac{V_m}{V}\right)^n \quad \text{...} \quad [4.2.25]$$

The subscript m here denotes the maximum elevation condition.

Actual gas pressure is now:

$$p = p_a - 101300 \, \text{Pa}$$

The gas force on the piston is:

$$P = pA$$

The force available for balancing the out of balance moment is:

$$P_e = P - F_s \quad \text{...} \quad [4.2.26]$$

F_s: seal friction force

The friction force due to friction at the bearing is calculated similarly as in the case of manual depression and applied to the force in the equation immediately above to determine the final balancing moment.

Out of balance moment and balancing moment curves during elevation and depression are shown in Fig 4.2.6 through the elevation range of a typical hydro-pneumatic balancing gear.

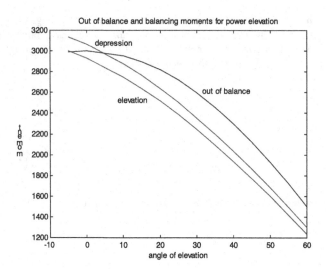

Fig 4.2.7: Plot of out of balance and balancing moments against elevation range for a typical hydro-pneumatic balancing gear.

4.3 Design of a Pneumatic Balancing Gear

Design Exercise 4.3.1

It is intended to design a pneumatic balancing gear for an artillery gun with the following characteristics:

Weight of elevating parts: 4000.00 N, the elevation range is from -5 to +60°. The dimensions a, b and c are selected as 0.75 m, 1.4 m and 0.5 m respectively. For equal load distribution, the arrangement consists of two identical pneumatic balancing gears placed at equal lateral distances from the bore centre line and below the level of the trunnions. The internal diameter of the pistons is 0.095 m and that of the piston rods is 0.03 m. The effective piston area is 0.0128 m².

250

Solution

The first step in the design of the balancing gear is the computation of the balancing moment neglecting the effect of friction forces.

The length of the spring is calculated with the help of Equations [4.1.3] and [4.1.5]. The length of the moment arm of the balancing gear force is determined with the help of Equation [4.1.6].

The out of balance moment is found by applying Equation [4.1.1].

The actual gas pressures at zero and maximum elevations are found from Equations [4.2.1] and [4.2.3], by equating the out of balance moments to the gas force, which is the product of pressure and piston area, at zero and maximum elevation. The pressures so determined are reduced to their absolute values and the initial volume of the gas found by applying Boyle's Law in the form of Equation [4.2.5].

The balancing moment which is the product of gauge, or actual, pressure times the piston area is determined as follows. From the change in gas volume, at different angles of elevation, the absolute pressure is determined, again from Boyle's Law and converted to actual. The balancing moment is then found from Equation [4.2.6]. The entire computation is conveniently accomplished by construction of a worksheet constructed as given ahead:

Symbols Used in the Worksheets

D: piston diameter We: weight of elevating parts
d: rod diameter w: seal widths of piston and rod seals
A: piston area p0abs: balancing gear absolute pressure at zero elevation
nu: leakage factor pmabs: maximum absolute pressure in balancing gear
V0: initial gas volume Psp: seal spring pressure
Kp: Pressure Factor mus: seal friction coefficient

muB: coefficient of friction of bearings
rD=rC: bearing radiuses of cradle and bottom carriage attachments

	A	B	C	D	E	F	G	H	I
1	Worksheet to compute elevation torque for pneumatic balancing gears, effect of friction not considered								
2	b	1.40	a	0.75	D	0.095	A	0.0128	
3	c	0.50	We	4000.00	d	0.03			
4	p0abs	6.01E+05	pmabs	4.37E+05	V0	1.66E-02			
5	phi	phi(rad)	Sinphi	x	l	Mob	deltax	P	p
6	-5	-0.09	-0.09	1.53	0.46	2988.58	-0.04	6544.69	512797.53
7	0	0.00	0.00	1.49	0.47	3000.00	0.00	6371.17	499201.76
8	10	0.17	0.17	1.40	0.49	2954.42	0.08	6010.54	470945.10
9	15	0.26	0.26	1.36	0.50	2897.78	0.13	5825.50	456446.85
10	20	0.35	0.34	1.32	0.50	2819.08	0.17	5638.89	441824.76
11	25	0.44	0.42	1.27	0.50	2718.92	0.21	5452.02	427183.14
12	30	0.52	0.50	1.23	0.49	2598.08	0.26	5266.37	412637.26
13	35	0.61	0.57	1.19	0.48	2457.46	0.30	5083.57	398314.33
14	40	0.70	0.64	1.14	0.47	2298.13	0.34	4905.41	384354.31
15	45	0.79	0.71	1.10	0.45	2121.32	0.38	4733.82	370910.28
16	50	0.87	0.77	1.07	0.42	1928.36	0.42	4570.95	358148.29
17	55	0.96	0.82	1.03	0.39	1720.73	0.46	4419.04	346246.05
18	60	1.05	0.87	1.00	0.35	1500.00	0.49	4280.49	335390.26

Fig 4.3.1: Worksheet to compute torque required to elevate without effect of friction

	J	K	L	M	N	O	P	Q
5	delataV	pa	V	pa	p	Pe	Mb	Te
6	-0.0005	614097.53	0.0161	619798.35	518498.35	6617.45	3021.81	-33.22
7	0.0000	600501.76	0.0166	600501.76	499201.76	6371.17	3000.00	0.00
8	0.0011	572245.10	0.0177	564006.90	462706.90	5905.40	2902.74	51.68
9	0.0016	557746.85	0.0182	546951.53	445651.53	5687.73	2829.24	68.53
10	0.0022	543124.76	0.0188	530764.32	429464.32	5481.13	2740.21	78.87
11	0.0027	528483.14	0.0193	515487.90	414187.90	5286.16	2636.21	82.71
12	0.0033	513937.26	0.0199	501157.93	399857.93	5103.27	2517.61	80.46
13	0.0038	499614.33	0.0204	487805.35	386505.35	4932.86	2384.60	72.86
14	0.0044	485654.31	0.0210	475458.46	374158.46	4775.28	2237.17	60.96
15	0.0049	472210.28	0.0215	464144.68	362844.68	4630.89	2075.19	46.13
16	0.0054	459448.29	0.0220	453892.02	352592.02	4500.03	1898.45	29.92
17	0.0058	447546.05	0.0224	444730.03	343430.03	4383.10	1706.73	13.99
18	0.0062	436690.26	0.0228	436690.26	335390.26	4280.49	1500.00	0.00

Fig 4.3.1(continued): Worksheet to compute torque required to elevate without effect
of friction

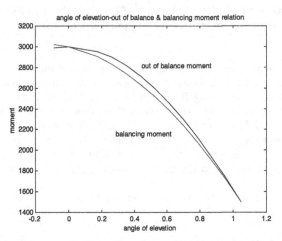

Fig 4.3.2: Graphical representation of angle of elevation-out of balance and balancing
moment without the effect of friction for the problem stated in Design Exercise 4.3.1

Manual Operation

In order to compute the actual torque necessary to operate the elevating gearing during elevation and depression, it is necessary to incorporate the effect of friction in the calculations. The friction effect is due, firstly, to the friction of the seals; secondly, to the friction in the bearings at the points of attachment of the balancing gear to the cradle and bottom carriage. The calculation is carried out separately for elevation and depression due to the change in direction of the friction force during the operations of elevation and depression.

Torque during Manual Elevation

The torque during elevation is calculated according to the following sequence:

Firstly the seal friction forces are determined. The friction force of the piston seal is found using Equation [4.2.7], wherein the pressure used is the actual pressure obtained from column O of the worksheet in Fig 4.3.1 and secondly for the piston rod seal using Equation [4.2.8] and the values of pressure as earlier stated. The sum of the friction forces due to the piston seal and the rod seal is then deducted from the balancing gear force obtained from column P of the worksheet in Fig 4.3.1. This force now becomes the balancing gear force for the purposes of calculation of friction force at the bearings.

Friction forces at the attachment bearings are computed as follows. The friction force at the bottom carriage bearing is computed using Equation [4.2.13] and the friction force at the cradle bearing using Equation [4.2.14]. The sum of these two friction forces, vide equation [4.2.15] are then subtracted from the balancing gear force to give the final balancing force as per Equation [4.2.19]. This balancing force multiplied by the balancing gear force moment arm yields the balancing moment as in Equation [4.2.20]. The difference between the out of balance moment and the balancing moment is the torque required to elevate the elevating parts. Equation [4.2.21] refers.

The calculations are shown in the worksheet of Fig 4.3.3.

	A	B	C	D	E	F	G	H	I	J	K	L	M	N	O	P
1	Worksheet to compute torque for manual elevation, including effect of friction: Pneumatic balancing gears															
2	nu	0.88	Psp	105369.35	mus	0.05	D	0.095	rD=rC	0.0127						
3	Kp	0.73	w	0.015	c	0.50	d	0.03	muB	0.08						
4	Seal friction calculations			Bearing friction calculations									Torque calculations			
5	phi	fp	fr	fp+fr	Pe	x	l	d	TD	TC	FD	FC	Pb	Mb	Mob	Te
6	-5	101.94	32.19	134.13	6483.32	1.53	0.46	0.68	6.59	6.59	-8.03	-6.40	6497.74	2967.14	2988.58	21.44
7	0	98.79	31.20	129.98	6241.19	1.49	0.47	0.69	6.34	6.34	-7.24	-6.22	6254.65	2945.14	3000.00	54.86
8	10	92.82	29.31	122.14	5783.26	1.40	0.49	0.70	5.88	5.88	-5.98	-5.98	5795.21	2848.58	2954.42	105.84
9	15	90.04	28.43	118.47	5569.25	1.36	0.50	0.71	5.66	5.66	-5.47	-5.90	5580.63	2775.97	2897.78	121.81
10	20	87.39	27.60	114.99	5366.14	1.32	0.50	0.71	5.45	5.45	-5.04	-5.86	5377.05	2688.18	2819.08	130.90
11	25	84.90	26.81	111.71	5174.46	1.27	0.50	0.71	5.26	5.26	-4.69	-5.85	5185.00	2585.76	2718.92	133.16
12	30	82.56	26.07	108.63	4994.65	1.23	0.49	0.70	5.07	5.07	-4.41	-5.88	5004.94	2469.10	2598.08	128.98
13	35	80.37	25.38	105.75	4827.11	1.19	0.48	0.70	4.90	4.90	-4.20	-5.95	4837.25	2338.38	2457.46	119.08
14	40	78.36	24.74	103.10	4672.18	1.14	0.47	0.69	4.75	4.75	-4.07	-6.07	4682.31	2193.62	2298.13	104.52
15	45	76.51	24.16	100.67	4530.22	1.10	0.45	0.67	4.60	4.60	-4.03	-6.24	4540.49	2034.68	2121.32	86.64
16	50	74.83	23.63	98.46	4401.57	1.07	0.42	0.65	4.47	4.47	-4.10	-6.50	4412.17	1861.38	1928.36	66.98
17	55	73.33	23.16	96.49	4286.61	1.03	0.39	0.63	4.36	4.36	-4.31	-6.87	4297.79	1673.52	1720.73	47.21
18	60	72.02	22.74	94.76	4185.73	1.00	0.35	0.61	4.25	4.25	-4.72	-7.42	4197.86	1471.04	1500.00	28.96

Fig 4.3.3: Worksheet for calculation of torque, including friction effect for pneumatic balancing gear during elevation

Torque during Manual Depression

The calculations during manual depression are similar to that of manual elevation. However, during manual depression the seal friction force which is equal in magnitude to that in elevation changes in sign. Also the friction torques at the cradle and bottom carriage supports are also the same in magnitude to those in elevation, but different in sense. The differences in the calculation from those of manual elevation are given below.

The seal friction forces are determined as before but applied in the opposite sense to obtain the effective balancing gear force for the purposes of calculation of friction force at the bearings. Friction forces at the attachment bearings are computed as follows. The friction force at the bottom carriage bearing is computed using Equation [4.2.16] and the friction force at the cradle bearing using Equation [4.2.17]. The sum of these two friction forces, vide equation [4.2.18] are then subtracted from the balancing gear force to give the final balancing force as per Equation [4.2.19]. This balancing force multiplied by the balancing gear force moment arm yields the balancing moment as in Equation [4.2.20]. The difference between the out of balance moment and the balancing moment is the torque required to elevate the elevating parts. Equation [4.2.21] refers. The calculations are shown in the worksheet of Fig 4.3.4 ahead.

	A	B	C	D	E	F	G	H	I	J	K	L	M	N	O	P
1		Worksheet to compute torque required for depression, including effect of friction:														
2		nu	0.88	Psp	105369.35	mus	0.05	D	0.095	rD=rC	0.0127					
3		Kp	0.73	w		0.015	c	0.50	d	0.03	muB	0.08				
4		Seal friction calculations				Bearing friction calculations							Torque calculations			
5	phi	fp	fr	fp+fr	Pe	x	l	d	TD	TC	FD	FC	Fe	Mb	Mob	Te
6	-5	101.94	32.19	-134.13	6751.58	1.53	0.46	0.68	6.86	6.86	8.36	6.66	6736.56	3076.20	2988.58	-87.62
7	0	98.79	31.20	-129.98	6501.16	1.49	0.47	0.69	6.61	6.61	7.55	6.48	6487.13	3054.60	3000.00	-54.60
8	10	92.82	29.31	-122.14	6027.54	1.40	0.49	0.70	6.12	6.12	6.23	6.23	6015.08	2956.65	2954.42	-2.23
9	15	90.04	28.43	-118.47	5806.20	1.36	0.50	0.71	5.90	5.90	5.71	6.15	5794.34	2882.27	2897.78	15.50
10	20	87.39	27.60	-114.99	5596.12	1.32	0.50	0.71	5.69	5.69	5.26	6.11	5584.75	2792.01	2819.08	27.06
11	25	84.90	26.81	-111.71	5397.87	1.27	0.50	0.71	5.48	5.48	4.89	6.10	5386.87	2686.44	2718.92	32.49
12	30	82.56	26.07	-108.63	5211.90	1.23	0.49	0.70	5.30	5.30	4.60	6.14	5201.17	2565.91	2598.08	32.17
13	35	80.37	25.38	-105.75	5038.61	1.19	0.48	0.70	5.12	5.12	4.38	6.21	5028.02	2430.60	2457.46	26.85
14	40	78.36	24.74	-103.10	4878.38	1.14	0.47	0.69	4.96	4.96	4.25	6.33	4867.80	2280.52	2298.13	17.62
15	45	76.51	24.16	-100.67	4731.55	1.10	0.45	0.67	4.81	4.81	4.21	6.52	4720.82	2115.50	2121.32	5.83
16	50	74.83	23.63	-98.46	4598.50	1.07	0.42	0.65	4.67	4.67	4.28	6.79	4587.42	1935.31	1928.36	-6.95
17	55	73.33	23.16	-96.49	4479.59	1.03	0.39	0.63	4.55	4.55	4.50	7.18	4467.91	1739.76	1720.73	-19.03
18	60	72.02	22.74	-94.76	4375.26	1.00	0.35	0.61	4.45	4.45	4.93	7.75	4362.57	1528.76	1500.00	-28.76

Fig 4.3.4: Worksheet to calculate torque required for manual depression including the effect of friction

Consolidated graphical solution to Design Exercise 4.3.1 is contained in Fig 4.3.5, which shows the out of balance or weight moment and the balancing moments of the balancing gear during elevation and depression.

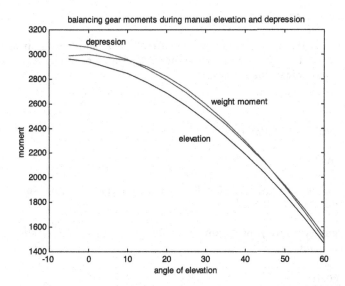

Fig 4.3.5: Balancing gear moments during manual elevation and depression including effect of friction

Design Exercise 4.3.2

Compute the balancing moments during elevation for the balancing gear as given in Design Exercise 4.3.1 for the case of power elevation and depression. Assume the value of the polytrophic exponent as 1.4.

Solution .

The problem is solved as follows:

Power elevation:

The absolute gas pressure at zero elevation is got from cell K7 of the worksheet in Fig 4.3.1. Applying Equation [4.2.22]:

The absolute gas pressure p of volume V of the gas at the angle of elevation under consideration is given by:

$$p_a = 600501.76 \left(\frac{0.0166}{V} \right)^n$$

The volume V of the gas at the angle of elevation under consideration is given by:

$$V = 0.0166 + A.x = 0.0166 + 0.0128x$$

Knowing x, the actual gas pressure can now be found from:

$$p = p_a - 101300 \text{ Pa}$$

The gas force on the piston is:

$$P = pA$$

The force available for balancing the out of balance moment is:

$$P_e = P - F_s, \quad F_s: \text{ seal friction force}$$

The friction force due to friction at the bearing is calculated similarly as in the case of manual elevation and applied to the force in the equation immediately above to determine the final balancing moment. The problem is conveniently solved by constructing a worksheet as in Fig 4.3.6 below:

	A	B	C	D	E	F	G	H	I	J
1			Worksheet to compute torque required for power elevation: pneumatic balancing gear							
2	nu	0.88	Psp	105369.35	mus	0.05	n	1.4	V0	0.0166
3	Kp	0.73	w	0.015	muB	0.08	p0abs	600501.7566	A	0.0128
4	c	0.50	rD=rC	0.0127			D	0.095	d	0.03
5			Gas pressure calculations				Seal friction calculations			
6	phi	x	dx	V	pa	p	P	fp	fr	fp+fr
7	-5	1.53	-0.04	0.0161	627689.51	526389.51	6718.16	103.23	32.60	135.83
8	0	1.49	0.00	0.0166	600501.76	499201.76	6371.17	98.79	31.20	129.98
9	10	1.40	0.08	0.0177	550037.72	448737.72	5727.11	90.54	28.59	119.13
10	15	1.36	0.13	0.0182	526893.29	425593.29	5431.73	86.76	27.40	114.16
11	20	1.32	0.17	0.0188	505192.26	403892.26	5154.76	83.21	26.28	109.49
12	25	1.27	0.21	0.0193	484953.56	383653.56	4896.46	79.91	25.23	105.14
13	30	1.23	0.26	0.0199	466185.47	364885.47	4656.93	76.84	24.27	101.11
14	35	1.19	0.30	0.0204	448889.52	347589.52	4436.19	74.01	23.37	97.39
15	40	1.14	0.34	0.0210	433063.81	331763.81	4234.21	71.43	22.56	93.98
16	45	1.10	0.38	0.0215	418705.84	317405.84	4050.96	69.08	21.82	90.90
17	50	1.07	0.42	0.0220	405814.78	304514.78	3886.44	66.98	21.15	88.13
18	55	1.03	0.46	0.0224	394393.11	293093.11	3740.67	65.11	20.56	85.67
19	60	1.00	0.49	0.0228	384447.64	283147.64	3613.73	63.48	20.05	83.53

Fig 4.3.6: Worksheet to compute balancing moment and power elevating torque for a pneumatic balancing gear

	A	B	C	D	E	F	G	H	I	J	K
21		Bearing friction calculations						Torque calculations			
22	Pe	l	d	TD	TC	FD	FC	Fe	Mb	Mob	Te
23	6582.33	0.46	0.68	6.69	6.69	8.15	6.49	6567.69	2999.08	2988.58	-10.50
24	6241.19	0.47	0.69	6.34	6.34	7.24	6.22	6227.72	2932.45	3000.00	67.55
25	5607.98	0.49	0.70	5.70	5.70	5.80	5.80	5596.39	2750.85	2954.42	203.57
26	5317.57	0.50	0.71	5.40	5.40	5.23	5.64	5306.71	2639.71	2897.78	258.06
27	5045.27	0.50	0.71	5.13	5.13	4.74	5.51	5035.02	2517.18	2819.08	301.89
28	4791.32	0.50	0.71	4.87	4.87	4.34	5.42	4781.56	2384.57	2718.92	334.36
29	4555.83	0.49	0.70	4.63	4.63	4.02	5.36	4546.44	2242.91	2598.08	355.17
30	4338.80	0.48	0.70	4.41	4.41	3.77	5.35	4329.68	2093.02	2457.46	364.44
31	4140.22	0.47	0.69	4.21	4.21	3.60	5.37	4131.25	1935.45	2298.13	362.69
32	3960.06	0.45	0.67	4.02	4.02	3.52	5.46	3951.09	1770.56	2121.32	350.76
33	3798.31	0.42	0.65	3.86	3.86	3.54	5.61	3789.16	1598.55	1928.36	329.81
34	3654.99	0.39	0.63	3.71	3.71	3.68	5.86	3645.46	1419.50	1720.73	301.23
35	3530.20	0.35	0.61	3.59	3.59	3.98	6.26	3519.97	1233.49	1500.00	266.51

Fig 4.3.6: (continued) Worksheet to compute balancing moment and power elevating torque for a pneumatic balancing gear

Balancing Moment during Power Depression

Computation of Gas Pressure

The absolute gas pressure during depression is computed applying Equation [4.2.25]:

$$p_a = 384447.64 \left(\frac{0.0166}{V} \right)^{1.4}$$

The volume V of the gas at the angle of elevation under consideration is given by:

$$V = 0.0166 + A.x = 0.0166 + 0.0128x$$

Actual gas pressure is now:

$$p = p_a - 101300 \, \text{Pa}$$

The gas force on the piston is:

$$P = pA$$

The force available for balancing the out of balance moment is:

$$P_e = P - F_s$$

The friction force due to friction at the bearing is calculated similarly as in the case of manual depression and applied to the force in the equation immediately above to determine the final balancing moment.

	A	B	C	D	E	F	G	H	I	J	K
1	Worksheet to compute torque required for power depression:										
2	nu	0.88	Psp	105369.35	mus	0.05	D		0.095	rD=rC	0.0127
3	Kp	0.73	w	0.015	c	0.50	d		0.03	muB	0.08
4	A	0.0128	p0abs	600501.7566	V0	0.0166	n		1.4		
5	Gas pressure calculations							Seal friction calculations			
6	phi	x	dx	V	pa	p	P	fp	fr	fp+fr	Pe
7	-5	1.53	-0.04	0.0161	627689.51	526389.51	6718.16	103.23	32.60	-135.83	6853.99
8	0	1.49	0.00	0.0166	600501.76	499201.76	6371.17	98.79	31.20	-129.98	6501.16
9	10	1.40	0.08	0.0177	550037.72	448737.72	5727.11	90.54	28.59	-119.13	5846.25
10	15	1.36	0.13	0.0182	526893.29	425593.29	5431.73	86.76	27.40	-114.16	5545.89
11	20	1.32	0.17	0.0188	505192.26	403892.26	5154.76	83.21	26.28	-109.49	5264.26
12	25	1.27	0.21	0.0193	484953.56	383653.56	4896.46	79.91	25.23	-105.14	5001.60
13	30	1.23	0.26	0.0199	466185.47	364885.47	4656.93	76.84	24.27	-101.11	4758.04
14	35	1.19	0.30	0.0204	448889.52	347589.52	4436.19	74.01	23.37	-97.39	4533.58
15	40	1.14	0.34	0.0210	433063.81	331763.81	4234.21	71.43	22.56	-93.98	4328.19
16	45	1.10	0.38	0.0215	418705.84	317405.84	4050.96	69.08	21.82	-90.90	4141.86
17	50	1.07	0.42	0.0220	405814.78	304514.78	3886.44	66.98	21.15	-88.13	3974.56
18	55	1.03	0.46	0.0224	394393.11	293093.11	3740.67	65.11	20.56	-85.67	3826.34
19	60	1.00	0.49	0.0228	384447.64	283147.64	3613.73	63.48	20.05	-83.53	3697.27
20											
21	Bearing friction calculations						Torque calculations				
22	l	d	TD	TC	FD	FC	Fe	Mb	Mob	Te	
23	0.46	0.68	6.96	6.96	-8.49	-6.76	6869.24	3136.79	2988.58	-148.20	
24	0.47	0.69	6.61	6.61	-7.55	-6.48	6515.18	3067.81	3000.00	-67.81	
25	0.49	0.70	5.94	5.94	-6.04	-6.04	5858.33	2879.61	2954.42	74.82	
26	0.50	0.71	5.63	5.63	-5.45	-5.88	5557.21	2764.32	2897.78	133.46	
27	0.50	0.71	5.35	5.35	-4.95	-5.75	5274.95	2637.14	2819.08	181.94	
28	0.50	0.71	5.08	5.08	-4.53	-5.66	5011.79	2499.38	2718.92	219.54	
29	0.49	0.70	4.83	4.83	-4.20	-5.60	4767.84	2352.13	2598.08	245.95	
30	0.48	0.70	4.61	4.61	-3.94	-5.59	4543.10	2196.19	2457.46	261.27	
31	0.47	0.69	4.40	4.40	-3.77	-5.62	4337.58	2032.11	2298.13	266.02	
32	0.45	0.67	4.21	4.21	-3.68	-5.71	4151.25	1860.26	2121.32	261.06	
33	0.42	0.65	4.04	4.04	-3.70	-5.87	3984.14	1680.80	1928.36	247.56	
34	0.39	0.63	3.89	3.89	-3.85	-6.14	3836.32	1493.82	1720.73	226.91	
35	0.35	0.61	3.76	3.76	-4.17	-6.55	3707.99	1299.38	1500.00	200.62	

Fig 4.3.7: Worksheet to compute the power torque for depression including the effect of friction

4.4 Spring Type Balancing Gears

Characteristics of Spring Type Balancing Gears

The operation of spring type balancing gears is characterized by the distinct advantage of independence from environmental conditions, especially temperature. Also they are unaffected by the rate of operation whether in elevation or depression. Pneumatic balancing gears, as earlier seen, are susceptible to variation in performance due to both these factors. Hence the design of spring type balancing gears is a much simpler process.

Design Process of Spring Type Balancing Gears

The design process of a spring type balancing gear follows three distinct stages:

Firstly is the determination of the optimum spring characteristics.

Secondly the torque required to elevate the elevating parts is determined which takes into account the friction at the points of attachment to the bottom carriage and cradle is estimated.

Finally, the torque requirement for depression is determined taking into account the friction force at the bearings.

Determination of Spring Characteristics

From Equation [4.1.1], the out of balance moment is:

$$M_{ob} = W_e a Cos\phi$$

The balancing moment is:

$$M_b = Pl$$

Here $P = P_0 - s(x_0 - x)$

In order to determine the spring stiffness factor s, and the initial spring force P_0, one of the methods introduced earlier may be employed.

The spring stiffness factor and the initial spring force having been determined, the wire diameter of the spring is found as follows:

By definition, the Spring Index of a helical spring is the ratio of mean diameter D to the wire diameter d:

$$C = \frac{D}{d} \quad\text{... [4.4.1]}$$

In a balancing gear, the mean diameter of the spring is usually defined by space constraints. Also the preferred range of the spring index is from 4 to 12. Spring index values of greater than 12 render the spring prone to buckling and spring index values of less then 4 pose manufacturing difficulties. The spring index value having been selected towards the centre of the preferred range, the wire diameter can be determined from Equation [4.4.1].

Maximum Shear Stress in the Spring

Two components of stress will act on any cross section of the spring coil. The first is a direct stress due to the spring force P, the second the torsional stress due to the torque T. The maximum shear stress in the coil, which is the direct sum of the two stresses and occurs at the innermost fibre is:

$$\tau_{max} = \frac{T_{max}r}{I} + \frac{P_{max}}{A} = \frac{P_m \dfrac{D}{2}\dfrac{d}{2}}{\dfrac{\pi}{32}d^4} + \frac{P_m}{\dfrac{\pi}{4}d^2}$$

Or:

$$\tau_{max} = \frac{8P_m D}{\pi d^3} + \frac{4P_m}{\pi d^2}$$

Substituting for the Spring Index C as given in Equation [4.4.1]:

$$\tau_{max} = \frac{8P_m C + 4P_m}{\pi d^2}$$

Or:

$$\tau_{max} = \frac{8P_m D}{\pi d^3}\left(1 + \frac{0.5}{C}\right)$$

Finally:

$$\tau_{max} = K_s \frac{8P_m D}{\pi d^3}$$ .... [4.4.2]

$$K_s = \left(1 + \frac{0.5}{C}\right)$$: Direct Shear Factor

The maximum shear stress having been determined from Equation [4.4.2], and given a Safety Factor of S, the maximum safe shear stress in the coil becomes:

$$\tau_s = S\tau_{max}$$ .... [4.4.3]

This value must not exceed the design stress for the wire of the calculated diameter.

Finally, the number of coils is determined from the relation:

$$N = \frac{Gd^4}{8sD^3}$$ .... [4.4.4]

The number of coils so determined is adjusted depending on the choice of end detail of the spring. For large diameter springs, as in balancing gears, this is usually done by squaring and grinding both ends. This reduces the number of active coils by 2, so the ultimate number of coils will be given by $N + 2$.

Determination of Elevation Torque Including Effect of Friction

In the case of spring type balancing gears the operations of manual and power traverse are identical. The Spring Stiffness Factor and the initial spring force having been established, the computations are fairly straightforward.

From the geometry and dimensions of the system, the spring length x at different increments of elevation angle is determined from Equation [4.1.5]:

$$x^2 = b^2 + c^2 - 2bcSin\phi$$

The moment arm l of the balancing gear force is found from Equation [4.1.6]:

$$l = \frac{bc}{x}Cos\phi$$

The distance d is found with the help of the relation:

$$d^2 = \sqrt{x^2 + l^2} \quad .... [4.4.5]$$

The spring force is calculated for increasing increments of elevation angle from the equation:

$$P = P_0 - s(x_0 - x)$$

From this spring force and knowing the radii of the bearings at the points of attachment of the balancing gear at the cradle and bottom carriage as also the coefficients of friction, the bearing torque is computed. Equations [4.2.11] and [4.2.12] refer.

The calculations for bearing friction force are the same as in the case of the pneumatic balancing gear and described by Equations [4.2.15] and [4.2.18].

The effective spring force P_b is obtained by subtracting the friction forces due to the friction torque from the spring force. The balancing moment is now given by:

$$M_b = P_b l$$

266

Finally the difference between the out of balance moment and the balancing moment is the torque necessary for elevation as in Equation [4.2.21]:

$$T_e = M_{ob} - M_b$$

Determination of Torque Required for Depression Including Effect of Friction

The computations to determine torque required during depression are identical to those of the elevation operation excepting for the calculation of the friction forces due to friction torque at the bearings.

Finally in the case of depression as in Equation [4.2.21a]:

$$T_d = M_{ob} - M_b$$

4.5 Design of a Spring Type Balancing Gear

Design Exercise 4.5.1

A spring type balancing gear is to be designed for an artillery gun. The following information and data is available at the outset:

The configuration of the system and dimensional data are contained in Fig [4.5.1]. The weight of the elevating parts is 4000 N. The coefficient of friction at the bearings is 0.08 and the radii of both the bottom carriage and cradle bearings are 0.0127 m. The elevation range of the gun is from -5° to +60°. Due to space limitations, the mean diameter of the spring is not to exceed 0.128 m. The material selected to be used is cold drawn steel wire. The maximum design stress against wire diameter curve for the given material appears in Fig 4.5.4. For effective load application the spring is to be squared and ground at both ends.

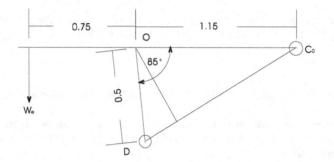

Fig 4.5.1: Configuration and dimensions of proposed balancing gear

Determine the optimum spring characteristics and hence compute the torque required during elevation and depression for the given configuration.

Determination of Spring Characteristics

The first step in the design process is determination of the spring stiffness factor and the assembled spring force. The calculations are contained in the worksheet of Fig 4.5.2 below.

Worksheet to compute spring displacement, Spring Stiffness Factor and assembled spring force:									
b	0.50	a	0.75	s	4285.70	theta0	1.48		
c	1.15	We	4000.00	P0	6371.20				

phi	phi(rad)	Cos(theta)	x	l	Mob	deltax	P	Mb	Te
-5	-0.09	0.00	1.25	0.46	2988.58	-0.04	6545.29	2989.83	-1.2415
0	0.00	0.09	1.21	0.47	3000.00	0.00	6371.20	3019.22	-19.2232
10	0.17	0.26	1.13	0.50	2954.42	0.08	6010.02	3014.15	-59.7229
15	0.26	0.34	1.09	0.51	2897.78	0.13	5824.89	2979.27	-81.4930
20	0.35	0.42	1.04	0.52	2819.08	0.17	5638.24	2922.70	-103.6258
25	0.44	0.50	1.00	0.52	2718.92	0.21	5451.39	2844.43	-125.5020
30	0.52	0.57	0.96	0.52	2598.08	0.26	5265.83	2744.46	-146.3831
35	0.61	0.64	0.91	0.52	2457.46	0.30	5083.25	2622.86	-165.4021
40	0.70	0.71	0.87	0.51	2298.13	0.34	4905.59	2479.69	-181.5611
45	0.79	0.77	0.83	0.49	2121.32	0.38	4735.02	2315.06	-193.7429
50	0.87	0.82	0.79	0.47	1928.36	0.42	4574.00	2129.11	-200.7463
55	0.96	0.87	0.76	0.43	1720.73	0.45	4425.28	1922.09	-201.3608
60	1.05	0.91	0.73	0.39	1500.00	0.49	4291.81	1694.49	-194.4938

Fig 4.5.2: Worksheet to determine spring displacement, Spring Stiffness Factor &
assembled spring force

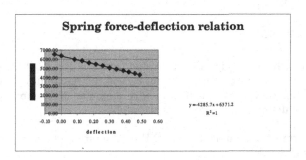

Fig 4.5.3: Assembled spring force and Spring Stiffness Factor by Trendline feature of
Excel

The Spring Index, chosen at the middle of the preferred range, is 8. From Equation
[4.4.1]:

269

$$d = \frac{D}{C} = \frac{0.128}{8} = 0.016 \text{ m or } 16.0 \text{ mm}$$

The Direct Shear Factor is:

$$K_s = 1 + \frac{0.5}{8} = 1.0625$$

The maximum spring force which occurs at the angle of maximum depression is:

$$P_m = P_0 - s(x_0 - x_m) = 6371.20 - 4285.70(1.21 - 1.25) = 6542.63 \text{ N}$$

The maximum shear stress in the spring from Equation [4.4.2] is:

$$\tau_{max} = K_s \frac{8.6542.63.0.128}{\pi.0.016^3} = 553.19 \text{ M Pa}$$

Taking into the given Safety Factor of 1.25, vide Equation [4.4.3], the safe shear stress in the wire becomes:

$$\tau_s = 1.25.553.19 = 691.5 \text{ M Pa}$$

On comparison with the empirically generated curves of Fig 4.5.4, it is seen that the maximum shear stress in the wire does not exceed the maximum shear stress of the material which for cold drawn steel, at a wire diameter of 16 mm, is 708.82 M Pa. Hence the spring diameter determined is acceptable.

It is left to determine the number of coils. From Equation [4.4.4] the number of coils in the spring is given by:

$$N = \frac{80.8.10^9.0.016^4}{8.4285.70.0.128^3} = 73.64 \approx 74$$

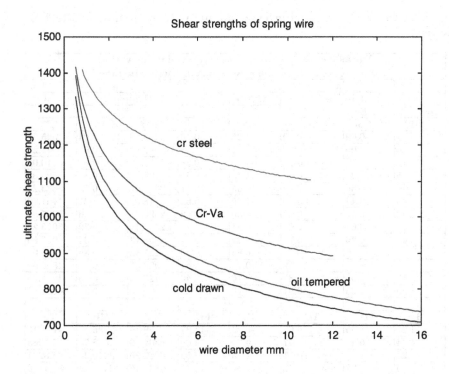

Fig 4.5.4: Ultimate shear strength of common spring alloys

Torque for Elevation for Spring Type balancing Gear Including Effect of Friction

Worksheet to compute torque required for elevation Spring type balancing gear, effect of friction included							
b	0.50	a		0.75	s	4285.70	theta0	1.48
c	1.15	We		4000.00	P0	6371.20	muB	0.08
rD=rC	1.27E-02							
phi	phi(rad)	Cos(theta)	x	l	Mob	deltax	P
-5	-0.09	0.00	1.25	0.46	2988.58	-0.04	6545.29
0	0.00	0.09	1.21	0.47	3000.00	0.00	6371.20
10	0.17	0.26	1.13	0.50	2954.42	0.08	6010.02
15	0.26	0.34	1.09	0.51	2897.78	0.13	5824.89
20	0.35	0.42	1.04	0.52	2819.08	0.17	5638.24
25	0.44	0.50	1.00	0.52	2718.92	0.21	5451.39
30	0.52	0.57	0.96	0.52	2598.08	0.26	5265.83
35	0.61	0.64	0.91	0.52	2457.46	0.30	5083.25
40	0.70	0.71	0.87	0.51	2298.13	0.34	4905.59
45	0.79	0.77	0.83	0.49	2121.32	0.38	4735.02
50	0.87	0.82	0.79	0.47	1928.36	0.42	4574.00
55	0.96	0.87	0.76	0.43	1720.73	0.45	4425.28
60	1.05	0.91	0.73	0.39	1500.00	0.49	4291.81

d	TD	TC	FD	FC	Pb	Mb	Te
1.24	6.65	6.65	-0.19	-14.37	6559.85	2996.48	-7.8915
1.24	6.47	6.47	0.34	-14.00	6384.86	3025.70	-25.6963
1.25	6.11	6.11	1.35	-13.53	6022.20	3020.25	-65.8291
1.26	5.92	5.92	1.84	-13.41	5836.46	2985.19	-87.4110
1.26	5.73	5.73	2.32	-13.37	5649.29	2928.43	-109.3542
1.26	5.54	5.54	2.81	-13.42	5462.01	2849.96	-131.0407
1.26	5.35	5.35	3.30	-13.57	5276.10	2749.81	-151.7332
1.26	5.16	5.16	3.81	-13.82	5093.26	2628.02	-170.5667
1.26	4.98	4.98	4.35	-14.21	4915.45	2484.68	-186.5452
1.25	4.81	4.81	4.95	-14.79	4744.86	2319.87	-198.5537
1.24	4.65	4.65	5.62	-15.60	4583.99	2133.76	-205.3935
1.23	4.50	4.50	6.41	-16.76	4435.63	1926.59	-205.8569
1.22	4.36	4.36	7.40	-18.44	4302.86	1698.85	-198.8543

Fig 4.5.5: Worksheet to compute torque required for elevation: Spring type balancing gear

Torque for Depression for Spring Type balancing Gear Including Effect of Friction

Worksheet to compute torque required for depression							
Spring type balancing gear, effect of friction included							
b	0.50	a	0.75	s	4285.70	theta0	1.48353
c	1.15	We	4000.00	P0	6371.20	muB	0.08
rD=rC	1.27E-02						
phi	phi(rad)	Cos(theta	x	l	Mob	deltax	P
-5	-0.09	0.00	1.25	0.46	2988.58	-0.04	6545.29
0	0.00	0.09	1.21	0.47	3000.00	0.00	6371.20
10	0.17	0.26	1.13	0.50	2954.42	0.08	6010.02
15	0.26	0.34	1.09	0.51	2897.78	0.13	5824.89
20	0.35	0.42	1.04	0.52	2819.08	0.17	5638.24
25	0.44	0.50	1.00	0.52	2718.92	0.21	5451.39
30	0.52	0.57	0.96	0.52	2598.08	0.26	5265.83
35	0.61	0.64	0.91	0.52	2457.46	0.30	5083.25
40	0.70	0.71	0.87	0.51	2298.13	0.34	4905.59
45	0.79	0.77	0.83	0.49	2121.32	0.38	4735.02
50	0.87	0.82	0.79	0.47	1928.36	0.42	4574.00
55	0.96	0.87	0.76	0.43	1720.73	0.45	4425.28
60	1.05	0.91	0.73	0.39	1500.00	0.49	4291.81

d	TD	TC	FD	FC	Pb	Mb	Td
1.24	6.65	6.65	0.19	14.37	6530.74	2983.18	5.41
1.24	6.47	6.47	-0.34	14.00	6357.54	3012.75	-12.75
1.25	6.11	6.11	-1.35	13.53	5997.84	3008.04	-53.62
1.26	5.92	5.92	-1.84	13.41	5813.32	2973.35	-75.57
1.26	5.73	5.73	-2.32	13.37	5627.19	2916.98	-97.90
1.26	5.54	5.54	-2.81	13.42	5440.78	2838.89	-119.96
1.26	5.35	5.35	-3.30	13.57	5255.57	2739.11	-141.03
1.26	5.16	5.16	-3.81	13.82	5073.24	2617.69	-160.24
1.26	4.98	4.98	-4.35	14.21	4895.73	2474.71	-176.58
1.25	4.81	4.81	-4.95	14.79	4725.18	2310.25	-188.93
1.24	4.65	4.65	-5.62	15.60	4564.02	2124.46	-196.10
1.23	4.50	4.50	-6.41	16.76	4414.93	1917.59	-196.86
1.22	4.36	4.36	-7.40	18.44	4280.77	1690.13	-190.13

Fig 4.5.6: Worksheet to compute torque required for depression: Spring type balancing gear

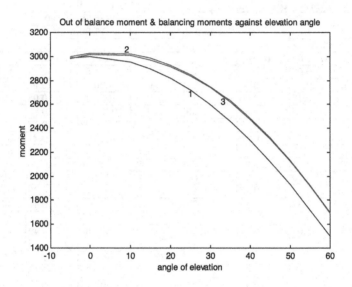

Fig 4.5.7: Out of balance & balancing moments against angle of elevation for the spring type balancing gear of Design Exercise 4.5.1.

Curves as indicated are:

1. Out of balance moment.
2. Balancing moment during elevation.
3. Balancing moment during depression.

5

Design of Elevating Mechanisms

5.1 Design Considerations of Elevating Mechanisms

Elevating Mechanisms in General

A weapon is aimed at a target by imparting to it a vertical angle, the angle of elevation and a horizontal angle; the angle in azimuth. Elevating mechanisms are necessary when effort required for the rotation of the elevating parts about the trunnion axis, in order to impart the necessary vertical ballistic angle between the axis of the bore and the horizontal is beyond the convenient capability of the average person responsible for aiming the weapon. To ensure accuracy, the elevating mechanism must be extremely sensitive and precise. Sensitivity means that the elevating mechanism should facilitate the rotation of the weapon about the trunnion axis to stop and start exactly when required to do so by the operator. Precise here implies the movement of the weapon must be in accordance with its accuracy requirement. Further, once aimed, the weapon must be held at the angle of elevation selected by the firer until deliberately altered.

Types of Elevating Mechanisms

Depending upon the tactical role of the weapon, elevating mechanisms are broadly divided into manually operated and power driven mechanisms. Clearly, this nomenclature is derived from the nature of the power source. In the case of power driven mechanisms, manual operation is invariably provided as a back up. Power driven mechanisms may again be electrical or hydraulic. Electrically driven systems suffer from unavoidable power losses as a result of power conversion from mechanical to electrical at the power source and back from electrical to mechanical at the point of application. In the case of hydraulic systems, high pressure lines are a source of peril, if ruptured, especially in the closed confines of fighting compartments.

Components of an Elevating Mechanism

Gear Train

The gear train in an elevating mechanism serves to transform the large torque, about trunnion axis required to elevate the elevating parts, to a manageable torque at the operating hand wheel. The gear train in conjunction with shafts and joints allows the torque applied at the gunner's station, to be transferred to the point of application at the elevating arc with the least possible friction loss.

The selection of gear ratio depends on the torque and speed requirements at the elevating gear and the nature of the power source. In the case of manual operation this torque should not exceed 5.5 N-m for ease of operation. The gear ratio is also selected so that one complete rotation of the hand wheel results in an easily related angular response of the elevating parts. For instance, one turn of the hand wheel corresponding to a gear ratio of 360 will result in a change in elevation of the bore axis by 1º.

In the case of fast moving targets, the speed of the priority target dictates the speed of elevation This speed of elevation may only be attainable with power elevation. For tanks and anti aircraft weapons, which are designed to track and fire on moving targets, the elevating speed is of the order of 60º per second. The gear ratio in this case becomes the ratio of the elevating speed desired to the motor speed. For convenience, motors with standard speeds are used and the gear ratio is selected to conform to this requirement.

Having established the overall gear ratio in keeping with the constraints mentioned above, the gear train itself is outlined to connect the elevating arc with the hand wheel, based on their locations and separation. The types of gears comprising a gear train are selected depending on the speed of operation, the direction of transfer of motion from shaft to shaft, the torque ratios between shafts and space considerations as also from tooth wear and heat generation aspects. Helical gears are preferred for high-speed operation. Worm gears are used when shafts are non-parallel and high torque ratios are the requirement. Worm pairs also possess the distinct advantage of self-locking, if so designed. However they suffer from heating generating problems due to inherent sliding motion. Bevel gears are suitable for shafts at 90°. Generally, the gear train should consist of the fewest members to minimize friction and inertia losses. Conversely, gear trains should consist of light and compact components to minimize torques necessary to drive the train itself and to achieve overall compactness. Backlash in the train is also an important consideration. For the train as a whole, it should not exceed around 3 minutes to fulfill the requirement of high accuracy.

Locks

Locks are devices, which serve two important functions. Firstly they hold the elevating parts at the desired angle of elevation and they prevent inadvertent rotation of the elevating parts during firing when high torques act about the trunnions. Secondly they counteract reverse rotation being transmitted from the elevating arc to the hand wheel or power source, which would endanger the operator. Locks are of three types; worm pairs designed to self lock, one way clutches and brakes. Of the three only brakes have the capability to stop bi-directional movement.

Worm pairs are favored for elevating mechanisms of equipments, which are manually operated and are placed preferably near the elevating arc so that they lock out the reverse load on the rest of the gear train.

Brakes find application in AFVs, which have to contend with reverse torques, which change in direction as a result of pitching of the vehicle. Brakes may be inserted at any intermediate position of the gear train, the closer to the elevating arc; the greater has to be the capacity of the brake.

Clutches are used to regulate the quantity of torque being transmitted especially during the application of a load. Slipping clutches are used to minimize the effects of large reverse torques especially in tanks.

Buffers

Buffers are devices used to bring the elevating parts, which are being rapidly elevated or depressed to a smooth and controlled halt. With manually operated weapons, contact between metal lugs on the top carriage and cradle suffice to limit the movement. Filled in teeth on the elevating arc also serve the same purpose. With power elevation, hydraulic buffers are preferred which brake the movement of the elevating parts by resistance offered to the flow of liquid through constrictions.

Stabilizers

Combat vehicles such as tanks and ICVs are designed to track and fire on targets on the move. Stabilizers are intended to maintain the attitude of the gun regardless of the pitching and yawing movements of the vehicle. The stabilizer consists essentially of a reference system, sensors that define the attitude of the gun with respect to this reference system and servos, which activate the elevating gearing to compensate for changes in the attitude of the gun as a result of vehicular pitching and yawing. Stabilized elevating systems are therefore more sophisticated versions of basic elevating mechanisms.

Balancing Gears

Balancing gears are important to elevating mechanisms in that they diminish the magnitude of the torque required to elevate the gun as also offer resistance during the action of depression, thereby reducing the power requirement at the source to within practical limits. Efficient balancing is therefore a prerequisite to successful elevating mechanism design.

5.2 Elevating Gear Trains

Torque Calculations

The method of torque computation and the static and dynamic loads on the elevating gear have been dealt with in Reference 1. The relevant equations are repeated below for ease of reference.

Static Loads on Elevating Gears

The unbalanced moment about the trunnions is given by:

$$T_e = M_{ob} - M_b \quad\text{..}\quad [6.2.1]\ \text{Reference 1}$$

M_{ob}: weight moment
M_b: balancing gear moment

The static friction moment in the trunnion bearings is given by:

$$T_b = \mu F_t r_b \quad\text{..}\quad [6.2.2]\ \text{Reference 1}$$

F_t: load on both trunnions
r_b: trunnion bearing radius
μ: coefficient of friction.

Dynamic Loads on Elevating Gears

The firing couple which acts on the elevating gearing is given by:

$$T_f = aP - bF_B \quad\text{...}\quad [6.2.3]\ \text{Reference 1}$$

P: propellant gas force
F_B: braking force of recoil system

Elevating Torque

The torque required to accelerate the elevating parts and the gear train itself is:

$$T_{aE} = \left(I_S \frac{r_E}{r_P} \frac{r_G}{r_S} + \frac{1}{\eta} \frac{r_G}{r_S} \frac{r_E}{r_P} I_A + \frac{1}{\eta^2} \frac{r_G}{r_S} \frac{r_E}{r_P} I_E \right) \alpha_E \quad \text{..........................[6.2.9] Reference 1}$$

I: mass moment of inertia of the shaft about the trunnion axis
α: angular acceleration of the elevating parts
r: gear pitch radius
Subscripts S, A, E, G, P denote individual shafts/gears

The net torque at the elevating arc is given by:

$$T_E = T_{aE} + T_b + T_f + T_e \quad \text{...[6.2.5] Reference 1}$$

Manual Elevation

For a gun, which is elevated manually and does not fire while being elevated, the acceleration of the elevating parts is nonexistent and the firing couple is also zero. The relevant expression is:

$$T_E = T_b + T_e \quad \text{..[6.2.6] Reference 1}$$

Power Elevation

For a gun which is elevated by power elevation and which does not fire during the process of elevation:

$$T_E = T_{aE} + T_b + T_e \quad \text{...[6.2.7] Reference 1}$$

The torque at the hand wheel or power source required for elevation is:

$$T_S = \frac{1}{\eta^n} \frac{T_E}{G.R.} \quad \text{..[6.2.8] Reference 1}$$

G.R.: overall gear ratio.

T_S: torque at source or hand wheel
n: number of gear meshes

Definitions Associated with Gears

Pitch circle: The circumference of an imaginary cylinder, which rolls without slipping when in friction contact with another cylinder.

Circular Pitch: The arc length on the pitch circle of the gear between successive tooth profiles.

Pressure Angle: The acute angle between the common tangent to the teeth profiles and the line of centres of the meshing gears.

Diametral Pitch: Defined as number of teeth per inch of pitch diameter, used only with the English system of units.

$$p_d = \frac{n}{d_p} \text{ teeth/in} \dotfill [5.2.1]$$

n: number of teeth
d_p: pitch diameter

Module: The pitch diameter in mm divided by the number of teeth used with SI or metric units.

$$m = \frac{d_p}{n} \text{ mm/tooth} \dotfill [5.2.2]$$

Gear Tooth Limiting Loads

Bending and surface fatigue are established by comparing the calculated total gear tooth tangential load with the maximum permissible values of tooth load based on bending and surface fatigue strengths. The maximum load on the tooth should not exceed either the bending fatigue limiting load or the surface fatigue-limiting load. Expressed symbolically:

$$\dot{F}_T \leq F_s \text{ and } F_T \leq F_W$$

F_T: maximum tooth load; F_S: maximum bending fatigue load or strength capacity
F_W: surface fatigue limiting load or wear capacity

Lewis Equation for Gear Tooth Bending Stress Analysis

With reference to Fig 5.2.1, for a uniform stress cantilever of constant width b, the thickness t, must vary parabolically with x. This parabolic cantilever is used as the basic model for analyzing the bending stress of gear teeth.

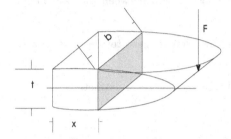

Fig 5.2.1: Constant stress cantilever

The Lewis Equation is based on the following simplifying assumptions:

The full load is applied at the tooth tip.
The radial component of the force applied is negligible.
The applied load is distributed uniformly across the entire face width of the tooth.
Sliding friction forces are negligible.
Stress concentrations are non-existent.

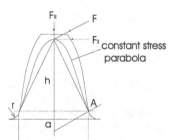

Fig 5.2.2: Comparison of bending stress in a gear tooth with a constant stress parabola

From Fig 5.2.2, the strength of the gear tooth exceeds that defined by the constant strength parabola, except at the point A where the strengths are equal. The tooth face width, indicated in Fig 5.2.1, is b.

The bending stress at point A:

$$\sigma = \frac{Mt}{I} = \frac{6F_s h}{bt^2} \quad \text{.. [5.2.3]}$$

The terms $\frac{6h}{t^2}$ are related purely to the geometry of the tooth and may be written as a function of circular pitch:

$$p_c = \frac{t^2}{6h} \quad \text{or} \quad y = \frac{t^2}{6hp_c}$$

y is a dimensionless constant depending on the number of teeth and the tooth geometry and is called the Lewis Form Factor.

Substituting for y in Equation [5.2.3], the Lewis expression for maximum permissible force from bending stress consideration becomes:

$$F_s = \sigma p_c by \quad \text{... [5.2.4]}$$

When using SI units the form of the Lewis expression changes to:

$$F_s = \sigma mbY \quad \text{.. [5.2.4a]}$$

Here, the following substitutions made are: $p_c = \pi m$ and $Y = \pi y$. It must be noted that in the above equation, if b and m in mm and σ in MPa, the load results in N. The values of the Lewis Form Factor Y are reproduced in Fig 5.2.3 for spur gears and for worms in Fig 5.2.4.

No of	Pressure angle		
teeth	20° stub teeth	20° full length teeth	25°
12	0.33512	0.22960	0.27677
13	0.34827	0.24317	0.29281
14	0.35985	0.25530	0.30717
15	0.37013	0.26622	0.32009
16	0.37931	0.27610	0.33178
17	0.38757	0.28508	0.34240
18	0.39502	0.29327	0.35210
19	0.40179	0.30078	0.36099
20	0.40797	0.30769	0.36916
21	0.41363	0.31406	0.37671
22	0.41883	0.31997	0.38370
24	0.42806	0.33056	0.39624
26	0.43601	0.33979	0.40717
28	0.44294	0.34790	0.41678
30	0.44902	0.35510	0.42530
34	0.45920	0.36731	0.43976
38	0.46740	0.37727	0.45156
45	0.47846	0.39093	0.46774
50	0.48458	0.39860	0.47681
60	0.49391	0.41047	0.49086
75	0.50345	0.42283	0.50546
100	0.51321	0.43574	0.52071
150	0.52321	0.44930	0.53668
300	0.53348	0.46364	0.55351
rack	0.54406	0.47897	0.57139

Fig 5.2.3: Lewis Form Factor Y for spur gears

Pressure angle degrees	Lewis Form Factor Y
14.5	0.3142
20	0.3927
25	0.4712
30	0.5498

Fig 5.2.4: Lewis Form Factor for worm gears

From the Lewis equation it can be concluded that bending stresses in a gear tooth vary directly with load and inversely with tooth width, circular pitch analogous to size and inversely according to the Lewis Form Factor.

Barth's Velocity Factor

When gears and pinions in mesh are driven at speed they are subject to dynamic loading which is a function of the speed, the greater the speed, the higher the dynamic loading. The estimation of effect of dynamic loading was based on comparison of the failure of identical gears under static and dynamic loading. A factor known as the Barth's Velocity Factor, named after the person who first propounded it, at a specified pitch line velocity is given by the Barth Equation:

$$K_v = \frac{3}{3 + v_p} \quad\text{..} [5.2.5]$$

The Barth Equation for teeth cut or milled but not precisely generated was subsequently modified to:

$$K_v = \frac{6}{6 + v_p} \quad\text{..} [5.2.6]$$

Currently the AGMA advocates the following formulae for non-precision spur gears:

$$K_v = \frac{50}{50 + (200 v_p)^{0.5}} \quad\text{..} [5.2.7]$$

In the case of precision gears:

$$K_v = \left[\frac{78}{78 + (200 v_p)^{0.5}} \right]^{0.5} \quad\text{..} [5.2.8]$$

v_p: pitch line velocity in m/s

Introducing the velocity factor, Equations [5.2.4] and [5.2.4a] become:
In the US specification system:

$$F_s = K_v \varpi p_c by \dotfill [5.2.9]$$

In the SI system:

$$F_s = K_v \varpi mbY \dotfill [5.2.10]$$

Spur Gear Trains

Basic Tooth Dimensions

Basic tooth dimensions for full depth teeth possessing pressure angles of 20° and 25° are given below in the SI system of units:

Dimension	Relation	
	20°	25°
Minimum No of teeth on pinion	18	12
Minimum total No of teeth	36	24
Addendum	m	m
Dedendum	1.25m	m
Tooth thickness	m/2	m
Minimum width of top land	0.25m	0.25m

Fig 5.2.5: Basic spur gear tooth dimensions in the SI system

Standard Modules

In order to minimize the inventory of gear cutting tools, standard modules of 1, 1.25, 1.5, 2, 2.5, 3, 4, 5, 6, 8, 10, 12, 16, 20, 25, 32, 40 and 50 are used:

The standards given above are for ease of manufacture. In the case of the design of high performance gears the standards given may serve as a guideline and not a restriction.

Surface Fatigue Limiting Load or Wear Capacity

The wear capacity of a gear is a function of the material, curvature, length of line of contact and smoothness of the sliding surfaces in contact. The following empirical equation suffices for rough estimates, assuming effective lubrication:

$$F_W = d_p b K_W \dotfill [5.2.11]$$

d_p: pitch diameter
K_W: empirical Wear Factor representing material and tooth geometry

Material	Hardness BHN	Bending stress MPa spur	bevel	K_w MPa 20°
Cast iron	180	34.5	20.7	0.524
Cast iron	230	48.3	27.6	0.759
Steel	180	137.9	69.0	0.407
Steel	230	151.7	75.9	0.552
Steel	325	220.7	103.4	1.124
Flame hardened		241.4	103.4	2.876
Case hardened		379.3	206.9	4.490

Fig 5.2.6: Wear Factors for spur and bevel gears based on BHN; 20° pressure angle

Material		K_W					
Worm	Worm gear	$\lambda < 10°$		$\lambda < 25°$		$\lambda > 25°$	
		psi	N/mm²	psi	N/mm²	psi	N/mm²
Steel 250 BHN	Sand cast bronze	60	0.4137	75	0.517125	90	0.62055
Face hardened steel 500BHN	Sand cast bronze	80	0.5516	100	0.68950	120	0.82740
Cast iron	Sand cast bronze	150	1.03425	185	1.275575	225	0.1.551375

Fig 5.2.7: Worm pair Wear Factors for different common materials and lead angles

287

Estimation of Gear Dimensions

It is necessary to establish the following parameters of a spur gear in order to calculate the tooth loads:

Pitch diameter:

$$d_p = \frac{mn}{10^3} \, \text{m} \quad\text{..} \quad [5.2.12]$$

Pitch line velocity:

$$v_p = \frac{\pi d_p N}{60} \, \text{m/s} \quad\text{...} \quad [5.2.13]$$

Load on gear tooth:

$$F_T = \frac{W}{v_p} \quad\text{...} \quad [5.2.14]$$

Velocity factor is selected from Equations [5.2.6] to [5.2.8]

Width of tooth face:

$$b = \frac{F_T}{K_v m Y \sigma} \quad\text{...} \quad [5.2.15]$$

N: speed in RPM
W: power transmitted in watts

The face width of the tooth is dictated by load considerations. If the tooth is too wide, the load distribution across the tooth face will not be uniform. In order to ensure uniform load distribution across a wide tooth face the finishing, bearings and rigidity of the assembly has to meet stringent requirements. For normal purposes the tooth face width is fixed by selection of the module in Equation [5.2.15] above so that the tooth face width in mm satisfies the following condition:

$3\pi m \geq b \geq 5\pi m$

Worm Pairs

A worm pair consists of a worm also called worm pinion and a worm gear, also known as a worm wheel. The worm is a helical gear with a large helix angle so that a single tooth wraps around its surface continuously. The geometry of a worm gear is similar to that of a power screw with the worm taking the place of the screw thread and the worm gear serving as its nut. Worms usually have only one thread and can hence create gear ratios as large as the number of teeth on the worm gear, this within a compact space. Normally worm gears are made of a special bronze SAE 65 for which a value of maximum bending stress of 24000 psi has been established.

Definitions Associated with Worm Gears

Lead: The distance that a point on the worm gear moves axially in one revolution of the worm.

$L = \pi m n_w$, m being the module and n_w the number of teeth on the worm.

Lead angle: The lead divided by the circumference of the pitch circle of diameter d_{pw} of the worm is the tangent of the lead angle λ :

$$Tan\lambda = \frac{L}{\pi d_{pw}} \quad \text{.. [5.2.16]}$$

Axial pitch: The axial pitch p_{aw} of a worm equals the circular pitch p_{cG} of the worm gear and is related to the lead by the number of worm gear teeth:

$$p_{aw} = p_{cG} = \frac{\pi d_{pG}}{n_G} \quad \text{.. [5.2.17]}$$

Addendum of a worm tooth: $a = 0.3183 p_a$

Dedendum of a worm tooth: $c = 0.3683 p_a$

Pitch diameter of the worm: There is a limitation on the minimum pitch diameter of a shell worm due to strength considerations. Worms cut onto shafts can have lesser pitch diameters than shell worms. From experience the minimum pitch diameter of a worm is given by:

$$d_{pw} = 2.4\pi m + 27.94 \text{ mm} \quad\text{...} \quad [5.2.18]$$

Face width of the worm gear: According to AGMA the maximum face width of the worm gear is related to the worm pitch diameter d_{pw} by the expression:

$$b \leq 0.67 d_{pw} \quad\text{...} \quad [5.2.19]$$

Efficiency of a Worm Gear

The efficiency equation for a worm and worm gear pair is based on the efficiency analysis of a power screw. Keeping in mind that the worm lead angle corresponds to the screw lead angle and the thread angle of a screw thread, in the plane normal to the teeth, corresponds to the pitch angle of the worm, it is possible to apply the force, self locking and efficiency equation of a power screw to a worm and worm gear pair.

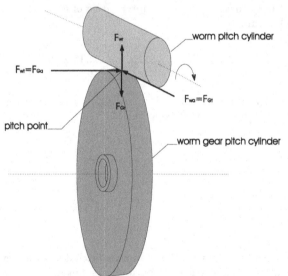

Fig 5.2.8: Worm and worm gear force diagram; worm driving in direction as indicated

F: force

Subscripts:

w: worm
G: worm gear
a: axial
t: tangential
r: radial

The relationship between worm and worm gear forces for the usual 90° shaft angle is depicted in Fig 5.2.7, for the worm driving in the direction indicated. In this case the worm tangential force equals the worm gear axial force. It follows that the worm axial force equals the worm gear tangential force. The worm radial force is equal in magnitude but opposite in direction to the worm gear radial force.

With reference to Fig 5.2.8, summing up the forces on the worm gear at the point of equilibrium:

In the axial direction of the worm:

$F_n Cos\beta Cos\lambda = \mu F_n Sin\lambda + R_b$

From which:

$$F_n = \frac{R_b}{Cos\beta Cos\lambda - \mu Sin\lambda} \quad\text{..}\quad [5.2.20a]$$

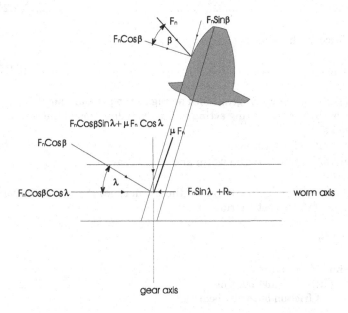

Fig 5.2.9: Forces on worm gear tooth with worm driving

In case friction is negligible:

$$F_n = \frac{R_b}{Cos\beta Cos\lambda}$$ .... [5.2.20b]

In the tangential direction of the worm:

$$F_{wt} = F_n Cos\beta Sin\lambda + \mu F_n Cos\lambda$$ .... [5.2.21a]

In case friction is nil:

$$F_{wt} = F_n Cos\beta Sin\lambda$$ .... [5.2.21b]

The radial force on the worm:

$$F_{wr} = F_n Sin\beta$$ .... [5.2.22]

F_n : normal force on the worm gear tooth acting at the pressure point.
R_b: reaction of the thrust bearing acting in the axial direction of the worm
λ : worm lead angle
β : pressure angle of the worm
μ : coefficient of friction between worm and worm gear

The torque now is given by the radius of the worm times the tangential force plus the friction torque of the thrust bearing:

$$T_f = r_w F_{wt} + \mu_b r_b R_b$$

r_w: pitch radius of the worm
r_b: effective radius of thrust bearing
μ_b : coefficient of friction of thrust bearing

Or:

$$T_f = r_w F_n \left(Cos\beta Sin\lambda + \mu Cos\lambda \right) + \mu_b r_b R_b$$

Substituting for F_n from Equation [5.2.20a]:

$$T_f = r_w \left(Cos\beta Sin\lambda + \mu Cos\lambda \right) \frac{R_b}{Cos\beta Cos\lambda - \mu Sin\lambda} + \mu_b r_b R_b$$

$$T_f = r_w R_b \left[\frac{\left(Cos\beta Sin\lambda + \mu Cos\lambda \right)}{\left(Cos\beta Cos\lambda - \mu Sin\lambda \right)} + \mu_b \frac{r_b}{r_w} \right]$$

In case friction in both the thread and the thrust bearing is eliminated, torque transmitted by the worm is:

$$T = r_w \, F_n Cos\beta Sin\lambda$$

Or:

$$T = r_w \, Cos\beta Sin\lambda \, \frac{R_b}{Cos\beta Cos\lambda} \, ; F_n \text{ is given by Equation [5.2.20b].}$$

Finally:

$$T = r_w R_b Tan\lambda$$

The efficiency of the worm is given by:

$$\eta = \frac{T}{T_f}$$

Or:

$$\eta = \frac{Tan\lambda}{\left(\dfrac{Cos\beta Sin\lambda + \mu Cos\lambda}{Cos\beta Cos\lambda - \mu Sin\lambda} \right) + \dfrac{r_b}{r_w} \mu_b} = \frac{Tan\lambda \left(Cos\beta Cos\lambda - \mu Sin\lambda \right)}{\left(Cos\beta Sin\lambda + \mu Cos\lambda \right) + \dfrac{r_b}{r_w} \mu_b \left(Cos\beta Cos\lambda - \mu Sin\lambda \right)}$$

$$= \frac{Sin\lambda(Cos\beta - \mu Tan\lambda)}{Cos\beta\left(Sin\lambda + \frac{r_b}{r_w}\mu_b Cos\lambda\right) + \mu Cos\lambda\left(1 - \frac{r_b}{r_w}\mu_b Tan\lambda\right)}$$

Finally the equation for efficiency of a worm pair taking both thread and thrust bearing friction into account is:

$$\eta = \frac{(Cos\beta - \mu Tan\lambda)}{Cos\beta\left(1 + \frac{r_b}{r_w}\mu_b Cot\lambda\right) + \mu Cot\lambda\left(1 - \frac{r_b}{r_w}\mu_b Tan\lambda\right)} \quad\quad\quad [5.2.23]$$

Self-Locking

This is an advantage over other types of gear sets, which finds application in armament gear trains. A self-locking worm set can only be driven from worm to worm gear thus locking out the high reverse torques occurring during firing. Self locking depends on the lead angle and the coefficient of friction which includes the surface finish, and lubrication

In the event of the gear driving the worm the contact shifts to the other side of the gear tooth and the normal force on the tooth reverses. The tangential force tending to drive the worm now is:

$$F_{wt} = F_n Cos\beta Sin\lambda - \mu F_n Cos\lambda \quad\quad\quad\quad\quad [5.2.24]$$

If this force becomes zero then as F_n is greater than zero:

$$Cos\beta Sin\lambda = \mu Cos\lambda$$

The condition for self-locking becomes:

$$\mu > Cos\beta Tan\lambda \quad\quad\quad\quad\quad\quad\quad\quad [5.2.25]$$

5.3 Design of a Manually Operated Elevating Gear Train

Design Exercise 5.3.1

Design of a Manual Operated Elevating Gear Train

It is required to design a manual elevating mechanism of the worm and worm gear type for an artillery field gun. The known data of the gun is detailed below.

Elevating torque after balancing is 3000.00 N-m. Load on the trunnions is 200.00 KN. Radius of the trunnion bearing is 50.8 mm. Coefficient of friction at the trunnions is 0.01. The maximum propellant gas force is 8 MN and the recoil brake force at the instant of maximum gas pressure is 5 MN, the moment arm of the gas force is 10 mm and the moment arm of the recoil brake force is 5 mm about the trunnion axis.

Only standard modules should be selected for the worm gear. Axial pitch of the worm is 15.875 mm. Number of threads in worm is one to achieve self-locking. Pressure angle of the worm is 20°. Coefficient of friction of the thread is 0.1. The material of the worm is cast iron and that of the worm gear phosphor bronze.

Solution

Torque due to friction at the trunnion bearing:

$$T_b = \mu r_b F_t = 0.01 . 0.0508 . 200000 = 101.6 \, \text{N-m}$$

Torque required for elevation, from Equation [6.2.6] Reference 1:

$$T_E = T_e + T_b = 3000.0 + 101.6 = 3101.6 \, \text{Nm}$$

For ease of manual operation the torque at the hand wheel should not exceed 5.65 N-m. Ignoring losses in the system the approximate gear ratio is given by:

$$G.R. = \frac{3101.6}{5.65} \approx 550$$

Relating the gear ratio to the elevation per turn of the hand wheel a one degree elevation will result for two turns of the hand wheel or half a degree for a single rotation of the hand wheel, if the gear ratio is 720, which is the nearest in terms of complete rotations of the hand wheel to the gear ratio of 550. Accepting for preliminary purposes a gear ratio of 720, the torque at the hand wheel becomes:

$$T_s = \frac{3101.6}{720} = 4.30 \, \text{N-m}, \text{ which is with the acceptable limit for manual operation.}$$

Gear Train

In order to proceed with the design of the elevating gear, it is decided that the gear train consist of a worm and worm gear, with a gear ratio of 72, shafted to a spur pinion which meshes with the elevating arc with a gear ratio of 10. To achieve the correct sense an idler may be interposed between elevating arc and spur pinion or the direction of threads on the worm reversed. The diagrammatic arrangement is shown in Fig 5.3.1.

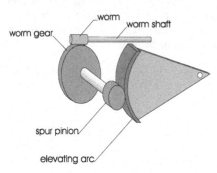

Fig 5.3.1: Worm based manual elevating mechanism (pitch cylinders depicted).

Self Locking

Pressure angle of the worm is 20°. Tangent of the lead angle is found from Equation [5.2.16]:

$$Tan\lambda = \frac{P_a}{\pi d_{pw}} = \frac{15.875}{\pi.65.64} = 0.0769$$

$Cos\beta Tan\lambda = 0.0719 < 0.1 = \mu$, hence it is confirmed that the condition for self locking is met.

Selection of Module

The module of the worm gear is given by:

$$m_G = \frac{d_{pG}}{n}$$

d_{pG}: pitch diameter of the worm gear
n: number of teeth of worm gear

Selecting module 5 as a representative example:

Pitch Diameter of Worm

From Equation [5.2.18], minimum pitch diameter of the worm for the selected module 5:

$$d_{pw} = 2.4\pi m + 27.94 = 65.64 \text{ mm}$$

Maximum Face Width of Worm Gear Tooth

The worms pitch diameter can be selected independent of the worm gear diameter. The magnitude of worm pitch diameter only affects the centre to centre distance between worm and worm gear and it not related to the gear ratio. However, according to Equation [5.2.19], the maximum face width of the worm gear according to AGMA is:

$$b = 0.67 d_{pw} = 0.67.65.64 = 43.98 \text{ mm}$$

Pitch Diameter of Worm Gear

Pitch diameter of the worm gear for selected standard module 5 from Equation [5.2.12] is:

$d_{pG} = m_G n = 5.72 = 360$ mm or 0.36 m

Maximum Tooth Load by Wear Consideration

Maximum tooth load for wear given as by Equation [5.2.11] is:

$$F_W = d_{pG}bK_w = 0.360.0.04398.1.034.10^6 = 16370.669 \text{ N}$$

Maximum Bending Load

The maximum bending stress for gear bronze has been established as 24 ksi or 165.5 M Pa. The Lewis Form factor Y obtained from Fig 5.2.4 for a worm gear is 0.3927.

Vide Equation [5.2.10], maximum tooth load from beam strength considerations:

$$F_S = \sigma_b mbY = 165.5.5.43.98.0.3927 = 14291.7 \text{ N}$$

Torque on the worm gear is:

$$T_G = \frac{T_E}{GR_{EG}} = \frac{3101.6}{10} = 310.16 \text{ N-m}$$

The maximum gear tooth load neglecting friction is:

$$F_T = \frac{T_G}{r_G} = \frac{310.6}{\dfrac{mn}{2000}} = 1725.56 \text{ N}$$

r_G : pitch radius of worm gear

Worksheet to determine optimum module: Elevating gear with worm pair						
n	72	Y	0.3927	TG Nm	310.16	
dpw m	0.04	TE Nm	3101.6	sigma		1.66E+08
Kw	1.03E+06	GR	10			
m mm	**dpw m**	**dpG m**	**bG m**	**FW N**	**FS N**	**FT N**
2	0.0430	0.144	0.029	4292.69	3746.54	4307.78
2.5	0.0468	0.180	0.031	5836.09	5093.57	3446.22
3	0.0506	0.216	0.034	7567.57	6604.77	2871.85
4	0.0581	0.288	0.039	11594.81	10119.63	2153.89
5	0.0656	0.360	0.044	16374.41	14291.12	1723.11
6	0.0732	0.432	0.049	21906.36	19119.26	1435.93
8	0.0883	0.576	0.059	35227.33	30745.43	1076.94
10	0.1033	0.720	0.069	51557.73	44998.15	861.56
12	0.1184	0.864	0.079	70897.56	61877.42	717.96
16	0.1486	1.152	0.100	118605.51	103515.58	538.47
20	0.1787	1.440	0.120	178351.16	155659.93	430.78
25	0.2164	1.800	0.145	269961.26	235614.67	344.62
32	0.2692	2.304	0.180	429814.40	375130.03	269.24
40	0.3295	2.880	0.221	657645.11	573974.33	215.39
50	0.4049	3.600	0.271	1010145.62	881626.95	172.31

The maximum tooth load must be less than the maximum tooth wear and bending loads. Also the spread between the maximum wear and the bending loads and the maximum tooth load must be within reasonable limits. In worksheet of Fig 5.3.2, the loads are computed for different standard modules. It is determined that the best condition is achieved for a standard module of 2.5.

Fig 5.3.2: Selection of optimum module based on maximum wear load, bending load and maximum tooth load

The design procedure for the spur pinion and elevating arc follows standard procedure and is detailed in the following design exercise, hence not elaborated further here.

5.4 Design of a Power Operated Elevating Gear Train

Design Exercise 5.4.1

Estimate the leading characteristics of a suitable spur gear train for an electrically powered elevating gear train for a heavy artillery gun. The data initially available is given as follows. Weight of the elevating parts is 60 KN. Load on the trunnions is 200 KN. The out of balance moment after balancing is 3000 Nm. Mass moment of inertia of the elevating parts about the trunnion axis is 236170.0 Kg-m². Mass density of gear steel is 7700 Kg/m³. The stipulated elevating acceleration is 6°/s² and the elevating velocity must not exceed 30°/s. Radius of the trunnion bearings is 50.8mm and the coefficient of friction at the bearing is 0.01. Efficiency at each meshing is 0.98. The rated speed of the electric driving motor is 1500 RPM. Also estimate the horsepower of the motor.

Solution

Torque Equations

The equations for torque related to elevating mechanisms from Reference 1 are given at the beginning of Section 2 of this chapter.

Acceleration Torque

Elevating acceleration is 6°/s² or 0.1 radians/s².

Vide Equation [6.2.4] Reference 1, torque required to accelerate the elevating parts:

$$T_{aE} = I_E \alpha_E = 236170.0.0.1 = 23617.0 \text{ Nm}$$

Vide Equation [6.2.2] Reference 1, static friction force moment in the bearings is:

$$T_b = \mu r_b F_t = 0.01.0.0508.200000 = 101.6 \text{ Nm}$$

Applying Equation [6.2.7] Reference 1, the torque requirement at the elevating arc is:

$$T_E = T_{\alpha E} + T_b + T_e = 23617.0 + 101.6 + 3000 = 26718.60 \text{ Nm}$$

The overall gear ratio is determined from the motor speed and the stipulated elevating speed.

The stipulated maximum elevating speed is 30°/s or 5 RPM

$$GR = \frac{\omega_m}{\omega_e} = \frac{1500}{5} = 300$$

A number of combinations are possible to suit the geometry of the layout. The gear train is here selected to comprise of 4 gear pairs as depicted in Fig 5.4.1. Because of the large radius of the elevating arc the first reduction is large. The subsequent reductions are fairly equal to have almost similar size gears and a compact gear train. The gear ratios selected for the purpose of this exercise are given below:

Gear Ratios

Elevating arc to pinion No 1: 10
Gear No 2 to pinion No 3:   5
Gear no 4 to pinion No 5:   3
Gear no 6 to pinion No 7:   2

When the pinion and gear are of identical material, the pinion teeth are weaker than the gear teeth because the teeth of the smaller gear are subject to greater undercutting. Hence the stress analysis is based on pinion dimensions. The calculations proceed as follows:

Torque

Torque at the elevating gear is obtained from Equation [6.2.7], Reference 1, which in this case is 26718.60 Nm. For the subsequent gear shaft which connects pinion No 1 and gear No 2 the torque is obtained from Equation [6.2.8], Reference 1:

$$Ts = \frac{1}{\eta^n} \frac{T_E}{G.R.} \text{ or } T_{1-2} = \frac{1}{0.98} \frac{26718.60}{10} = 2726.38 \text{ Nm}$$

Similarly, knowing the gear ratio and the torque in the preceding shaft, the torque can be calculated down the train to the motor shaft.

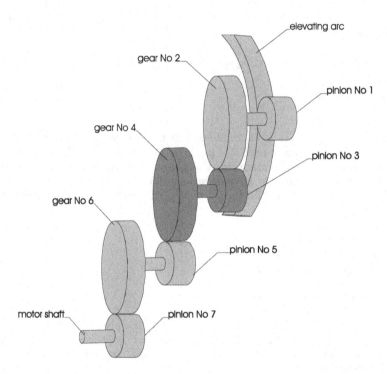

Fig 5.4.1: Position of gears and pinions in proposed gear train design (pitch cylinders depicted)

Power transmitted

Power transmitted by the shaft is calculated using the well-known formula:

$W = \dfrac{2\pi NT}{60}$ Watts, here N is the shaft speed in RPM, T the torque in N-m.

Pitch Diameter

Pitch diameter is calculated from Equation [5.2.12]

Pitch Line Velocity

Pitch line velocity is calculated with the help of Equation [5.2.13].

Module

Standard values of module are arbitrarily chosen as 2, 2.5, 3, 4, 5, 6, 8, 10,12, 16, 20, 25, 32,40 and 50. Based on the results of the calculations, additional module values may need to be considered.

Tooth load

Tooth load is given by Equation [5.2.14].

Barth's Velocity Factor

The velocity factor is calculated using Equation [5.2.6].

Lewis Form Factor

The Lewis factor is obtained from Fig 5.2.3, for the number of teeth under consideration and 20° pressure angle.

Tooth Face Width

The tooth face width is obtained from Equation [5.2.15].

Minimum and Maximum Tooth Face Widths

Minimum and maximum tooth face widths are established using the relation expressed by the condition $3\pi m \geq b \geq 5\pi m$.

Selection of tooth face width

The computation usually results in multiple options of module and number of teeth. As a guide a gear with a greater number of teeth is preferable from undercutting point of view and from overall compactness considerations, a smaller module is desirable.

The calculations are given in the worksheets of Figs 5.4.2 to 5.4.5. Computations in the worksheet are based on varying whole numbers of teeth for standard modules beginning with the minimum of 18 teeth which is the lowest number of teeth necessary to minimize undercutting.

Worksheet for determining face width and number of teeth					
Elevating arc and pinion 1					
Gear ratio	10	Elev arc speed RPM	5	sigmap MPa	345
Min teeth on pinion	18	Torque Nm	26718.6	Pinion No 1 speed RPM	50
Min teeth on gear	180	Power tx W	13989.83		

Pinion teeth	Arc teeth	Pitch diameter m						Velocity m/s					
		2.5	3	4	5	6	8	2.5	3	4	5	6	8
18	180	0.045	0.054	0.072	0.090	0.108	0.144	0.12	0.14	0.19	0.24	0.28	0.38
24	240	0.060	0.072	0.096	0.120	0.144	0.192	0.16	0.19	0.25	0.31	0.38	0.50
30	300	0.075	0.090	0.120	0.150	0.180	0.240	0.20	0.24	0.31	0.39	0.47	0.63
36	360	0.090	0.108	0.144	0.180	0.216	0.288	0.24	0.28	0.38	0.47	0.57	0.75
42	420	0.105	0.126	0.168	0.210	0.252	0.336	0.27	0.33	0.44	0.55	0.66	0.88
48	480	0.120	0.144	0.192	0.240	0.288	0.384	0.31	0.38	0.50	0.63	0.75	1.01
54	540	0.135	0.162	0.216	0.270	0.324	0.432	0.35	0.42	0.57	0.71	0.85	1.13

Pinion teeth	Arc teeth	Velocity factor						Transmitted load KN					
		2.5	3	4	5	6	8	2.5	3	4	5	6	8
18	180	0.98	0.98	0.97	0.96	0.95	0.94	118.75	98.96	74.22	59.37	49.48	37.11
24	240	0.97	0.97	0.96	0.95	0.94	0.92	89.06	74.22	55.66	44.53	37.11	27.83
30	300	0.97	0.96	0.95	0.94	0.93	0.91	71.25	59.37	44.53	35.62	29.69	22.27
36	360	0.96	0.95	0.94	0.93	0.91	0.89	59.37	49.48	37.11	29.69	24.74	18.55
42	420	0.96	0.95	0.93	0.92	0.90	0.87	50.89	42.41	31.81	25.45	21.21	15.90
48	480	0.95	0.94	0.92	0.91	0.89	0.86	44.53	37.11	27.83	22.27	18.55	13.92
54	540	0.94	0.93	0.91	0.89	0.88	0.84	39.58	32.99	24.74	19.79	16.49	12.37

Pinion teeth	Arc teeth	Lewis Factor Y	Tooth width b mm				
			2.5	3	4	5	6	8
18	180	0.29327	478.68	333.70	189.15	121.98	85.35	48.73
24	240	0.33056	320.56	223.75	127.13	82.18	57.64	33.06
30	300	0.35510	240.25	167.90	95.63	61.96	43.56	25.10
36	360	0.37229	192.17	134.46	76.77	49.86	35.13	20.33
42	420	0.38703	159.44	111.70	63.92	41.61	29.38	17.07
48	480	0.39553	137.37	96.34	55.26	36.05	25.51	14.88
54	540	0.40335	120.48	84.60	48.63	31.80	22.55	13.21
Minimum face width			23.56	28.27	37.70	47.12	56.55	75.40
Maximum face width			39.27	47.12	62.83	78.54	94.25	125.66

Fig 5.4.2: Worksheet to compute tooth face width of pinion No 1

Worksheet for determining face width and number of teeth					
Gear 2 and pinion 3					
Gear ratio	5	Gear 2 speed RPM	50	sigmap MPa	345
Min teeth on pinion	18	Torque	2726.39	Pinion speed RPM	250
Min teeth on gear	90	Power tx W	14275.33		

Pinion teeth	Gear teeth	Pitch diameter m						Velocity m/s					
		2.5	3	4	5	6	8	2.5	3	4	5	6	8
18	90.00	0.045	0.054	0.072	0.090	0.108	0.144	0.59	0.71	0.94	1.18	1.41	1.88
24	120.00	0.060	0.072	0.096	0.120	0.144	0.192	0.79	0.94	1.26	1.57	1.88	2.51
30	150.00	0.075	0.090	0.120	0.150	0.180	0.240	0.98	1.18	1.57	1.96	2.36	3.14
36	180.00	0.090	0.108	0.144	0.180	0.216	0.288	1.18	1.41	1.88	2.36	2.83	3.77
42	210.00	0.105	0.126	0.168	0.210	0.252	0.336	1.37	1.65	2.20	2.75	3.30	4.40
48	240.00	0.120	0.144	0.192	0.240	0.288	0.384	1.57	1.88	2.51	3.14	3.77	5.03
54	270.00	0.135	0.162	0.216	0.270	0.324	0.432	1.77	2.12	2.83	3.53	4.24	5.65

Pinion teeth	Gear teeth	Velocity factor						Transmitted load KN					
		2.5	3	4	5	6	8	2.5	3	4	5	6	8
18	90.00	0.91	0.89	0.86	0.84	0.81	0.76	24.23	20.20	15.15	12.12	10.10	7.57
24	120.00	0.88	0.86	0.83	0.79	0.76	0.70	18.18	15.15	11.36	9.09	7.57	5.68
30	150.00	0.86	0.84	0.79	0.75	0.72	0.66	14.54	12.12	9.09	7.27	6.06	4.54
36	180.00	0.84	0.81	0.76	0.72	0.68	0.61	12.12	10.10	7.57	6.06	5.05	3.79
42	210.00	0.81	0.78	0.73	0.69	0.65	0.58	10.39	8.66	6.49	5.19	4.33	3.25
48	240.00	0.79	0.76	0.70	0.66	0.61	0.54	9.09	7.57	5.68	4.54	3.79	2.84
54	270.00	0.77	0.74	0.68	0.63	0.59	0.51	8.08	6.73	5.05	4.04	3.37	2.52

Pinion teeth	Gear teeth	LF Y	Tooth width b mm					
			2.5	3	4	5	6	8
18	90.00	0.29327	105.22	74.37	43.30	28.66	20.55	12.30
24	120.00	0.33056	72.10	51.23	30.12	20.11	14.54	8.83
30	150.00	0.35510	55.24	39.44	23.40	15.75	11.48	7.06
36	180.00	0.37229	45.15	32.38	19.37	13.14	9.64	6.00
42	210.00	0.38703	38.24	27.55	16.61	11.34	8.37	5.27
48	240.00	0.39553	33.61	24.31	14.76	10.15	7.53	4.78
54	270.00	0.40335	30.06	21.82	13.34	9.22	6.88	4.40
Minimum face width			23.56	28.27	37.70	47.12	56.55	75.40
Maximum face width			39.27	47.12	62.83	78.54	94.25	125.66

Fig 5.4.3: Worksheet to compute tooth face width of pinion No 3

Worksheet for determining face width and number of teeth					
			Gear 4 and pinion 5		
Gear ratio	3	Gear 4 speed	250	sigmap MPa	345
Min teeth on pinion	18	Torque	556.41	Pinion speed RPM	750
Min teeth on gear	54	Power tx	14566.67		

Pinion teeth	Gear teeth	Pitch diameter						Velocity					
		1.5	2	2.5	3	4	5	1.5	2	2.5	3	4	5
18	54.00	0.027	0.036	0.045	0.054	0.072	0.090	1.06	1.41	1.77	2.12	2.83	3.53
24	72.00	0.036	0.048	0.060	0.072	0.096	0.120	1.41	1.88	2.36	2.83	3.77	4.71
30	90.00	0.045	0.060	0.075	0.090	0.120	0.150	1.77	2.36	2.95	3.53	4.71	5.89
36	108.00	0.054	0.072	0.090	0.108	0.144	0.180	2.12	2.83	3.53	4.24	5.65	7.07
42	126.00	0.063	0.084	0.105	0.126	0.168	0.210	2.47	3.30	4.12	4.95	6.60	8.25
48	144.00	0.072	0.096	0.120	0.144	0.192	0.240	2.83	3.77	4.71	5.65	7.54	9.42
54	162.00	0.081	0.108	0.135	0.162	0.216	0.270	3.18	4.24	5.30	6.36	8.48	10.60

Pinion teeth	Gear teeth	Velocity factor						Transmitted load KN					
		1.5	2	2.5	3	4	5	2	2	2.5	3	4	5
18	54.00	0.85	0.81	0.77	0.74	0.68	0.63	13.74	10.30	8.24	6.87	5.15	4.12
24	72.00	0.81	0.76	0.72	0.68	0.61	0.56	10.30	7.73	6.18	5.15	3.86	3.09
30	90.00	0.77	0.72	0.67	0.63	0.56	0.50	8.24	6.18	4.95	4.12	3.09	2.47
36	108.00	0.74	0.68	0.63	0.59	0.51	0.46	6.87	5.15	4.12	3.43	2.58	2.06
42	126.00	0.71	0.65	0.59	0.55	0.48	0.42	5.89	4.42	3.53	2.94	2.21	1.77
48	144.00	0.68	0.61	0.56	0.51	0.44	0.39	5.15	3.86	3.09	2.58	1.93	1.55
54	162.00	0.65	0.59	0.53	0.49	0.41	0.36	4.58	3.43	2.75	2.29	1.72	1.37

Pinion teeth	Gear teeth	LF	Tooth width b					
		Y	1.5	2	2.5	3	4	5
18	54.00	0.29327	106.52	62.92	42.19	30.63	18.73	12.95
24	72.00	0.33056	74.43	44.53	30.20	22.15	13.79	9.68
30	90.00	0.35510	58.07	35.14	24.08	17.82	11.26	8.00
36	108.00	0.37229	48.26	29.51	20.40	15.21	9.74	6.99
42	126.00	0.38703	41.52	25.63	17.86	13.41	8.68	6.28
48	144.00	0.39553	37.03	23.05	16.18	12.22	7.99	5.82
54	162.00	0.40335	33.57	21.06	14.88	11.30	7.45	5.46
Minimum face width			14.14	18.85	23.56	28.27	37.70	47.12
Maximum face width			23.56	31.42	39.27	47.12	62.83	78.54

Fig 5.4.4: Worksheet to compute tooth face width of pinion No 5

Worksheet for determining face width and number of teeth					
		Gear 6 and pinion 7			
Gear ratio	2	Gear 6 speed	750	sigmap MPa	345
Min teeth on pinion	18	Torque	189.25	Pinion speed RPM	1500
Min teeth on gear	36	Power tx w	14863.95		

Pinion teeth	Gear teeth	Pitch diameter						Velocity					
		2.5	3	4	5	6	8	2.5	3	4	5	6	8
18	36.00	0.045	0.054	0.072	0.090	0.108	0.144	3.53	4.24	5.65	7.07	8.48	11.31
24	48.00	0.060	0.072	0.096	0.120	0.144	0.192	4.71	5.65	7.54	9.42	11.31	15.08
30	60.00	0.075	0.090	0.120	0.150	0.180	0.240	5.89	7.07	9.42	11.78	14.14	18.85
36	72.00	0.090	0.108	0.144	0.180	0.216	0.288	7.07	8.48	11.31	14.14	16.96	22.62
42	84.00	0.105	0.126	0.168	0.210	0.252	0.336	8.25	9.90	13.19	16.49	19.79	26.39
48	96.00	0.120	0.144	0.192	0.240	0.288	0.384	9.42	11.31	15.08	18.85	22.62	30.16
54	108.00	0.135	0.162	0.216	0.270	0.324	0.432	10.60	12.72	16.96	21.21	25.45	33.93

Pinion teeth	Gear teeth	Velocity factor						Transmitted load KN					
		2.5	3	4	5	6	8	2.5	3	4	5	6	8
18	36.00	0.63	0.59	0.51	0.46	0.41	0.35	4.21	3.50	2.63	2.10	1.75	1.31
24	48.00	0.56	0.51	0.44	0.39	0.35	0.28	3.15	2.63	1.97	1.58	1.31	0.99
30	60.00	0.50	0.46	0.39	0.34	0.30	0.24	2.52	2.10	1.58	1.26	1.05	0.79
36	72.00	0.46	0.41	0.35	0.30	0.26	0.21	2.10	1.75	1.31	1.05	0.88	0.66
42	84.00	0.42	0.38	0.31	0.27	0.23	0.19	1.80	1.50	1.13	0.90	0.75	0.56
48	96.00	0.39	0.35	0.28	0.24	0.21	0.17	1.58	1.31	0.99	0.79	0.66	0.49
54	108.00	0.36	0.32	0.26	0.22	0.19	0.15	1.40	1.17	0.88	0.70	0.58	0.44

Pinion teeth	Gear teeth	LF	Tooth width b					
		Y	2.5	3	4	5	6	8
18	36.00	0.29327	26.42	19.71	12.62	9.05	6.97	4.68
24	48.00	0.33056	19.75	14.92	9.75	7.11	5.54	3.80
30	60.00	0.35510	16.33	12.46	8.27	6.10	4.80	3.33
36	72.00	0.37229	14.26	10.98	7.38	5.49	4.35	3.05
42	84.00	0.38703	12.82	9.93	6.75	5.06	4.03	2.85
48	96.00	0.39553	11.88	9.26	6.34	4.79	3.83	2.72
54	108.00	0.40335	11.15	8.73	6.02	4.57	3.67	2.62
Minimum face width			23.56	28.27	37.70	47.12	56.55	75.40
Maximum face width			39.27	47.12	62.83	78.54	94.25	125.66

Fig 5.4.5: Worksheet to compute tooth face width of pinion No 7

Torque Required to Accelerate the Gear Train

The dimensions of the gears and pinions established it is possible now to compute the torque at source necessary to drive the gear train itself. The torque required to accelerate the elevating parts was determined earlier and included for in the calculations, hence the torque necessary to drive the train less the elevating parts is determined as follows:

The number of teeth corresponding to the modules selected. The tooth face widths and the shaft speeds are obtained from the worksheets of Figs 5.4.2 to 5.4.5. The pitch radii of the gears are obtained from Equation [5.2.12]. Given the mass density of the material of the gears, the mass of each gear is calculated. The mass moment of inertia of each gear is calculated treating the gear as a solid disc with thickness of the tooth face width and radius equal to the pitch radius from the formula:

$$I_G = \frac{1}{2}mr_p^2$$

The mass moment of inertia of the shaft is the sum of the mass moments of inertia of the gear and the pinion housed on the same shaft. The shaft torque is now the product of the mass moment of inertia and the angular acceleration of the shaft. The angular acceleration of the shaft is determined by applying the relevant gear ratios to the elevating acceleration down the train. The calculations for determination of the torque required to accelerate the gear train are given in the spread sheet of Fig 5.4.6 below.

Worksheet to compute torque for accelerating the gear train										
Mass density of steel Kg/m^3		7700.00	neta		0.98	alphaE	0.1			
Gear	Teeth	module	b m	rp m	m Kg	IG	Shaft	I	alpha	T Nm	Ts Nm
1	48	4.00	0.055	0.0960	12.320	0.056768					
2	240	2.50	0.034	0.3000	73.173	3.292791	1 to 2	3.349559	1	3.3496	106.76544
3	48	2.50	0.034	0.0600	2.927	0.005268					
4	72	2.50	0.030	0.0900	5.917	0.023966	3 to 4	0.029234	5	0.1462	0.91318
5	24	2.50	0.030	0.0300	0.657	0.000296					
6	36	2.50	0.026	0.0450	1.294	0.001310	5 to 6	0.001606	15	0.0241	0.04917
7	18	2.50	0.026	0.0225	0.324	0.000082	7	0.000082	30	0.0025	0.00246
Torque necessary to accelerate the gear train Nm										107.73025
Torque necessary to accelerate the gear train as a % of torque at source Nm										111.57053

Fig 5.4.6: Worksheet to compute torque necessary to accelerate the gear train

Estimation of Motor Horsepower

The torque in shaft connecting gear No 6 and pinion No 5 obtained from Fig 5.4.5 is 189.25 Nm. The torque in the motor shaft is:

$$T_s = \frac{1}{0.98}\frac{189.25}{2} = 96.56\,\text{Nm}$$

From the worksheet of Fig 5.4.6, it is established that the torque required to accelerate the gear train is 107.73 Nm. As a percentage of the torque at source it works out to 111.6 % and must be included the power calculations. Hence the power at source is:

$$W_s = \frac{2\pi 1500.204.3}{60} = 32 \text{ KW or } 43 \text{ HP}$$

6

Design of Traversing Mechanisms

6.1 General Considerations for Design of Traversing Mechanisms

Function and Components of Traversing Mechanisms

Traversing mechanisms of weapons are intended to rotate the axis of the bore about the traversing axis, to precisely the direction in azimuth desired by the firer. In equipments wherein the weight of the weapon precludes ordinary human effort a power source is used but invariably with manual back-up. Traversing mechanisms also serve to maintain the weapon in the direction desired by the firer until intentionally altered. Traversing mechanisms are designed based on the tactical employment of the weapon. The target characteristics dictate the precision and speed of operation of the mechanism. From the operating point of view, simplicity and ease of operation coupled with low power requirement is desirable. The traversing mechanism therefore is designed to provide the best possible solution between the requirements at the target and those at the operating end.

Traversing Parts

The traversing parts consist of the components of the weapon, which move, in traverse. This includes all the assemblies above the traversing bearing. In an artillery gun this would be comprised of the saddle, the cradle, gun complete with recoil system and barrel attachments, the balancing gear and the fire control system. With turret mounted guns the turret itself and the turret basket would be part of the traversing parts in addition to the components just mentioned in the case of an artillery gun.

Types of Traversing Mechanisms

Here two commonly encountered traversing mechanisms, firstly, the pivot type traversing and secondly the turret ring traversing mechanism, which affords all round traverse, will be discussed.

Pivot Traverse

The pivot traverse is commonly found on towed artillery and anti aircraft pieces. The heart of the mechanism is the pivot at the centre and bottom of the saddle, which rotates on top of a bearing on the top surface of the bottom carriage. This arrangement allows an arc of traverse limited only by stability considerations. The stability, in turn, is dictated by the geometry of the bottom carriage. If the bottom carriage is so designed, this arrangement is amenable to 360° traverse.

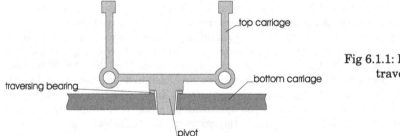

Fig 6.1.1: Pivot type traverse

Types of Pivot Traverse Mechanisms

The pivot type of traverse lends itself to different gearing arrangements. While infinite combinations based on different gear types are possible, the basic arrangements are described ahead:

Nut and Screw Traversing Mechanism

The nut and screw traversing mechanism consists of a nut pivoted in a recess in the bottom carriage through which a screw fixed by means of bearing to the top carriage is driven. The top carriage pivots about the bottom carriage by means of a central pivot, which is free to rotate in a cylindrical cavity in the bottom carriage. As the screw is driven into the nut, the top carriage to which it is attached is traversed about the pivot axis.

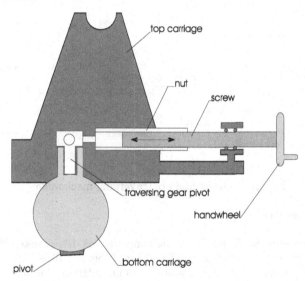

Fig 6.1.2: Nut and screw traversing mechanism

Pinion and Traversing Gear

The pinion and traversing arc consists of the traversing gear fixed horizontally to the top carriage, which is in mesh with a pinion driven by the operating hand wheel. In equipments, which are not designed for all round traverse, the traversing gear is replaced by a traversing arc corresponding to the angular range of traverse. The basic arrangement is depicted in Fig 6.1.3.

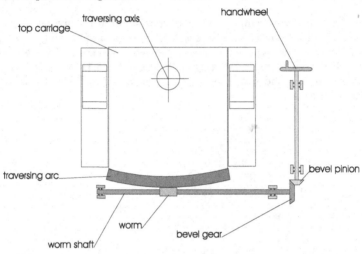

Fig 6.1.3: Pinion and arc traversing mechanism

Turret Ring Traverse

This type of traverse is found on vehicle mountings like tanks, ICVs and self propelled artillery. A ring spur gear with internal teeth is fitted inside the circular aperture on the topside of the hull. The turret is mounted on the hull with a circular bearing between the turret and hull. A pinion, mounted in bearings on the hull, meshes with the ring gear and when rotated by manual or powered effort transmitted through the gear train, causes the turret to rotate about the traversing axis. The traversing axis is the vertical through the centre of the pitch circle of the ring gear. A suitable gear train provides the necessary reduction. Brakes, clutches and buffers are employed to improve the performance of the system as a whole.

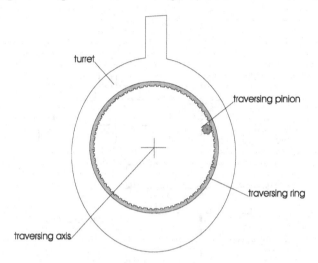

Fig 6.1.4: Turret ring traverse

Operation of a Typical Turret Traversing Mechanism

Line diagram of a typical turret traversing mechanism is shown in Fig 6.1.6 ahead. For manual operation, the annulus of the epicyclic gear train is unlocked by turning the selector upwards. During manual operation, the intermediate gear of the electric drive is locked by the thrust of the spring of the electromagnetic clutch. The sun of the epicyclic gear train is held. Motion is transmitted from the operating hand wheel to the worm pair, through the planetary carrier of the epicyclic gear train to the driving pinion, which runs over the teeth of the traversing ring in the hull causing the turret to rotate. For electrical operation, turning the handle downwards locks the annulus of the epicyclic gear train and renders the manual worm pair inoperative. It also switches on the electric power. When the electric power is switched on the electromagnetic clutch gets energized. The electromagnetic clutch now releases the intermediate cluster gear of the electric drive. Drive is transmitted through the

315

intermediate cluster gear to the friction clutch, to the axle, which mounts the sun of the epicyclic gear train to the planetary carrier to the drive pinion.

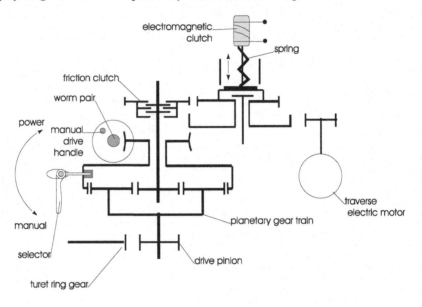

Fig 6.1.5: Line diagram of a turret traversing mechanism

6.2 Design of Traversing Gear Trains

Difference between Traversing and Elevating Mechanisms

From the design aspect, traversing mechanisms are closely related to elevating mechanisms. The main variation between the design process of the two lies in the estimation of the static and dynamic torques on the traversing arc, which are dissimilar from those acting on the elevating arc. Once this is computed the design methodology of the gear train follows similar procedures. For ease of reference, the torque equations of traversing mechanisms contained in Reference 1 are reproduced below.

Static Loads on the Traversing Mechanism

Weight Moment about the Traversing Axis

The weight moment about the traversing axis is given by:

$$M_T = W_t r Sin\,\theta Cos\,\psi \dots\dots\dots\dots\dots\dots\dots\dots [6.3.1]\ \text{Reference 1}$$

W_t: weight of traversing parts.
r: radius of traversing axis to CG of traversing mass.
θ: angle of inclination of ground
ψ: angle between the bore axis and the line through the axis of traverse at right angles to the direction of the slope of ground

Torque Due to Frictional Resistance in Traversing Bearing

$$T_b = \frac{2}{3}\mu F_n \frac{\left(r_o^3 - r_i^3\right)}{\left(r_o^2 - r_i^2\right)} \dots\dots\dots\dots\dots\dots\dots\dots [6.3.3]\ \text{Reference 1}$$

F_n: normal force on traversing bearing
r_i: inner radius of bearing
r_o: outer radius of bearing; μ: Coefficient of friction of bearing

Firing Couple due to the Eccentricity of the Recoiling Parts about the Traversing Axis

$$T_f = [aP - bF_B]Cos\,\phi \dots\dots\dots\dots\dots\dots\dots\dots [6.3.4]\ \text{Reference 1}$$

F_B: braking force of recoil system
P: propellant gas force
ϕ: angle of elevation.
a,b: moment arms of forces about traversing axis

Torque Required to Accelerate the Traversing Parts

$$T_{aT} = I_T \alpha_T \dots\dots\dots\dots\dots\dots\dots\dots [6.3.4a]\ \text{Reference 1}$$

I_T: mass moment of inertia of traversing parts about the traversing axis

a_T : angular acceleration of the traversing parts

Maximum Torque at the Traversing Gear

Maximum torque at the traversing gear is given by:

$$T_T = M_T + T_b + T_f + T_{aT} \quad\text{..[6.3.5] Reference 1}$$

Maximum Torque at the Traversing Gear under Different Conditions

Gun on level ground traversed manually, does not fire on the traverse:

$$T_T = T_b \quad\text{..[6.3.6] Reference 1}$$

In the condition that the same gun is on an incline:

$$T_T = M_T + T_b \quad\text{...[6.3.7] Reference 1}$$

Gun, with power traverse, on level ground, does not fire while traversing:

$$T_T = T_b + T_{aT} \quad\text{..[6.3.8] Reference 1}$$

For the same conditions, when the gun fires during traverse:

$$T_T = T_b + T_f + T_{aT} \quad\text{..[6.3.9] Reference 1}$$

For the same gun which is on a slope, does not fire while traversing:

$$T_T = M_T + T_b + T_{aT} \quad\text{..[6.3.10] Reference 1}$$

For the same gun which fires while traversing:

$$T_T = M_T + T_b + T_f + T_{aT} \quad\text{...[6.3.11] Reference 1}$$

Design of Traversing Gear Trains

The design of gear trains for traversing mechanisms follows the same procedure as for elevating gear trains in the case of spur and worm pairs. However in the case of traversing gear trains it is desirable to place the point of application of the traversing torque at the traversing arc at a point on the same vertical plane on which the axis of the bore lies and directly below it. This calls for transmission of motion between shafts intersecting at 90° for which straight bevel pairs are suitable.

Design of Bevel Gears for Traversing Gear Trains

Definitions Associated with Bevel Gears

Pitch cone: Cone containing the pitch elements of the teeth.

Cone centre: Point where the axes of the mating gears intersect.

Pitch angle: The smaller angle between the pitch line and the shaft axis.

Pitch diameter: The pitch of bevel gears is measured at the large end of the tooth and both the pitch diameter and the module are calculated as for spur gears.

Cone length: The length of the pitch cone element

$$l_c = \sqrt{\left(\frac{d_{pG}}{2}\right)^2 + \left(\frac{d_{pP}}{2}\right)^2} \quad \dots\dots\dots\dots\dots\dots\dots\dots\dots\dots\dots\dots\dots\dots\dots\dots\dots\dots\dots\dots\dots \text{[6.2.1]}$$

d_p is the pitch diameter, subscripts P and G imply pinion and gear respectively

Velocity ratio: Ratio of the pitch diameters of the bevel gear to the bevel pinion.

$$VR = \frac{d_{pG}}{d_{pP}} = \frac{N_P}{N_G} \quad \dots\dots\dots\dots\dots\dots\dots\dots\dots\dots\dots\dots\dots\dots\dots\dots\dots\dots\dots\dots\dots\dots \text{[6.2.2]}$$

319

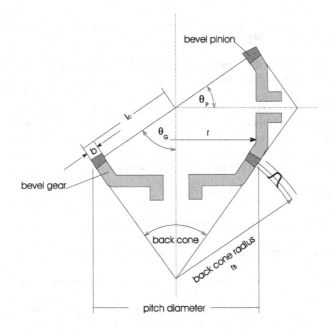

Fig 6.2.1: Bevel gear nomenclature

Determination of Pitch Angle for Bevel Gears

With reference to Fig 6.2.1:

When the shafts are at right angles the angle between the axis of the gear and the pinion is:

$$\theta_B = \frac{\pi}{2}$$

$$\theta_P = Tan^{-1} \frac{d_{pP}}{d_{pG}} = \frac{n_P}{n_G}$$ .... [6.2.3]

320

And:

$$\theta_G = Tan^{-1}\frac{d_{pG}}{d_{pP}} = \frac{n_G}{n_P} \quad\dots\dots\dots\dots\dots\dots\dots\dots\dots\dots\dots\dots\dots\dots\dots\dots\dots\dots\dots\dots\dots\dots [6.2.4]$$

Equivalent Number of Teeth of a Bevel Gear

This approximation, called Tregold's approximation, is based on the fact that a cone tangent to the sphere at the pitch point will closely approximate the sphere on either side of the pitch point for a short distance. The back cone may be developed as a plane surface and spur gear teeth corresponding to the pitch and pressure angle of the bevel gear and the radius of the developed cone can be drawn.

From the geometry of Fig 6.2.1:

$$Cos\theta = \frac{r}{r_B} \text{ or: } r_B = \frac{r}{Cos\theta}$$

The equivalent number of teeth is given by:

$$n_E = \frac{d_p^{'}}{m}, \ d_p^{'} \text{ is the back cone diameter and } m \text{ the module.}$$

$$n_E = \frac{2r}{mCos\theta} = \frac{T}{Cos\theta} \quad\dots\dots\dots\dots\dots\dots\dots\dots\dots\dots\dots\dots\dots\dots\dots\dots\dots\dots\dots\dots\dots [6.2.5]$$

T: actual number of teeth of pinion or gear as the case may be
θ : given by Equations [6.2.3] and [6.2.4] in the case of pinion and gear respectively

Strength of Bevel Gears

The strength of a bevel gear tooth is obtained using a modified form of the Lewis Equation:

$$F_S = \sigma K_v bmY'\frac{l_c - b}{l_c} \quad\dots\dots\dots\dots\dots\dots\dots\dots\dots\dots\dots\dots\dots\dots\dots\dots\dots\dots\dots [6.2.6]$$

Y': Lewis factor for the equivalent number of teeth

$\dfrac{l_c - b}{l_c}$: Bevel Factor. The cross section of bevel gear teeth reduces as the cone apex is approached; hence a bevel gear tooth is inherently weaker than a spur gear tooth of equal outer pitch diameter. Hence a correction factor called the Bevel Factor is incorporated in the strength equation.

Limiting Load for Wear

The limiting load for wear is obtained as for spur gears vide Equation [5.2.11], using the diameter of the back cone circle $d'_p = \dfrac{d_p}{Cos\theta}$:

$$F_W = d'_p b K_W Q' \ \dotfill \ [6.2.7]$$

The tooth face width b for a bevel gear should not exceed $\dfrac{1}{3} l_c$. The wear factor K_w for bevel gears may be obtained from the table in Fig 5.2.5. Q' is known as the Ratio Factor and is given by:

$$Q' = \dfrac{2 n_{EG}}{n_{EP} + n_{EG}} \ \dotfill \ [6.2.8]$$

Selection of Module

The module is selected based on the limiting strength loads for bending and wear. The limiting values of both these loads must be greater then the tangential tooth load. Further the differences between the limiting loads and the tooth load should not be excessive.

6.3 Design of a Manual Traversing Mechanism

Design Exercise 6.3.1

A manual traversing mechanism is to be designed for an artillery field gun. The design is to be based on the pivot type of traverse. Schematic of the proposed mechanism is as given in Fig 6.1.4. The traversing mechanism is to consist of a worm and traversing arc pair and a bevel gear and pinion as illustrated. Known data of the equipment is given below:

Weight of the traversing parts is 130 KN. The distance from the traversing axis to the centre of gravity of the traversing parts is 100 mm. The traversing mechanism is to cater for a maximum ground slope of 6°. The traversing range of the weapon is 45° either side of the centre line. The axis of the bore and the traversing axis lie on the same vertical plane.

The maximum normal firing load on the traverse bearing is estimated to be 250 KN. Dimensions of the traversing bearing are outer radius 0.2 m and inner radius 0.1 m. Coefficient of friction of the traversing bearing is 0.01.

Coefficient of friction of the worm thread is 0.1. Pressure angle of the worm is 20°. Effective radius of the worm thrust bearing is 0.05 m. Coefficient of friction of the worm thrust bearing is 0.02.

The bevel gear and pinion are both of steel of BHN 325 with a maximum bending stress of 103.4 M Pa.

Solution

Weight Moment

From Equation [6.3.1] Reference 1, the weight moment due to ground inclination and traverse is:

$$M_T = W_t r Sin\,\theta Cos\,\psi = 130000.0.1.0.1045284.0.7071 = 960.86 \text{ N-m}$$

Bearing Friction Torque

From Equation [6.3.3] Reference 1, torque due to frictional resistance in the traversing bearing is:

$$T_b = \frac{2}{3}\mu F_n \frac{\left(r_o^3 - r_i^3\right)}{\left(r_o^2 - r_i^2\right)} = \frac{2}{3}0.01.250000.\frac{0.2^3 - 0.1^3}{0.2^2 - 0.1^2} = 388.89 \text{ Nm}$$

For a manually traversed gun on an incline, vide Equation [6.3.7] Reference 1:

$T_T = M_T + T_b = 1349.75$ N-m

Since the hand wheel torque must not exceed 5.5 N-m for ease of operation, the overall gear ratio is:

$GR = \dfrac{1162.22}{5.65} = 238.89$. In terms of complete turn of the hand wheel the nearest convenient ratio is 360. So the torque at the source becomes 3.75 N-m, which is much more acceptable than 5.5 N-m.

Gear Ratios

The break down of the gear ratio is arbitrarily selected as 72 between the traversing arc and worm and 5 between the bevel gear and pinion. After the efficiency of the gear train is established, this ratio may be subject to some alteration.

Design of Worm-Traversing Arc Pair

The stresses in the traversing arc will always be higher than in the worm hence the stresses in the traversing gear arc form the basis of the design. The design procedure is as follows:

The chosen gear ratio between worm and traversing arc is 72. To achieve self locking, the worm will have a single start. To maintain the selected gear ratio, the traversing gear will have 72 teeth. The arc which will be a quadrant of 90º will have 22.5 or 23 teeth, 23 being the nearest whole number. It is now necessary to select a module from the range of standard modules, compute the bending stress load, the wear load and the maximum load on the tooth. The calculations are shown in the worksheet of Fig 6.3.2.

Module

Modules from 2 to 50 are selected from the standard range of modules. Sample calculations are given for arbitrarily selected module 4.

Pitch Diameter of Worm

The pitch diameter of the worm is calculated from Equation [5.2.18].

$d_{pw} = 2.4\pi m + 27.94$ mm $= 2.4.\pi.4 + 27.94 = 58.099$ mm

Pitch Diameter of the Traversing Arc

Pitch diameter of the traversing arc is computed from Equation [5.2.12]:

$d_{pG} = mn = 4.72 = 288$ mm

Tooth Face Width of Traversing Arc

According to Equation [5.2.19], the maximum face width of the worm gear according to AGMA is:

$b = 0.67 d_{pw} = 0.67.58.099 = 38.93$ mm

Maximum Tooth Load for Wear

Maximum tooth load for wear given as by Equation [5.2.11] is:

$F_W = d_{pG} b K_w = 0.288.0.03893.1.03425.10^6 = 11594.81$ N

Maximum Tooth Load from Bending Stress Consideration

The maximum tooth load from bending stress consideration is got from Equation [5.2.10]. For a manual traversing mechanism, as in this case, the velocity is deemed zero and the velocity factor K_v assumes a value of 1. The maximum bending stress of gear bronze has been established as 24 ksi or 165.5 MPa. Lewis Form Factor from Fig 5.2.4 is 0.3927

Maximum tooth load from beam strength considerations:

$F_S = \sigma_b m b Y = 165.5.4.38.93.0.3927 = 10120$ N

Maximum Tooth Load

For module 4, the pitch radius of the traversing arc is 144 mm. The maximum torque is given by Equation [6.3.7] Reference 1:

$$T_T = 1349.75 \ \text{N-m}$$

Hence the maximum tooth load is:

$$F_T = \frac{1349.75}{.144} = 9373.26 \ \text{N}$$

Selection of Module

The criteria for the selection of the module are that the maximum tooth wear and bending loads should both individually be greater than the maximum tooth load. Also the spread between the maximum tooth wear and bending loads and the maximum tooth load should be the minimum. From the results of the worksheet of Fig 6.3.2, it is seen that module 4 best meets this requirement.

Selection of module 4 is based on the loads as follows:

Maximum wear load: 11594.81 N
Maximum bending load: 10119.63 N
Maximum tooth load: 9373.26 N

Module 4 yields the following worm pair dimensions:

Pitch diameter of worm: 58.1 mm
Axial pitch of worm: 12.566 mm, from Equation [5.2.7]
Pitch diameter of traversing arc: 288 mm
Tooth face width of traversing arc: 39 mm

Worksheet to determine optimum module: Traversing gear with worm pair						
n	72	TT Nm	1349.75			
Kw	1.03E+06	GR	10			
Y	0.3927	sigma M Pa	1.66E+08			
m mm	dpw m	dpG m	bG m	FW N	FS N	FT N
2	0.0430	0.144	0.029	4292.69	3746.54	18746.53
2.5	0.0468	0.180	0.031	5836.09	5093.57	14997.22
3	0.0506	0.216	0.034	7567.57	6604.77	12497.69
4	0.0581	0.288	0.03893	11594.81	10119.63	9373.26
5	0.0656	0.360	0.044	16374.41	14291.12	7498.61
6	0.0732	0.432	0.049	21906.36	19119.26	6248.84
8	0.0883	0.576	0.059	35227.33	30745.43	4686.63
10	0.1033	0.720	0.069	51557.73	44998.15	3749.31
12	0.1184	0.864	0.079	70897.56	61877.42	3124.42
16	0.1486	1.152	0.100	118605.51	103515.58	2343.32
20	0.1787	1.440	0.120	178351.16	155659.93	1874.65
25	0.2164	1.800	0.145	269961.26	235614.67	1499.72
32	0.2692	2.304	0.180	429814.40	375130.03	1171.66
40	0.3295	2.880	0.221	657645.11	573974.33	937.33
50	0.4049	3.600	0.271	1010145.62	881626.95	749.86

Fig 6.3.1: Worksheet for computation of tooth loads and selection of optimum standard module for worm pair

Self-Locking

Pressure angle of the worm is 20°. Tangent of the lead angle is found from Equation [5.2.16]:

$$Tan\lambda = \frac{L}{\pi d_{pw}} = \frac{12.566}{\pi.58.1} = 0.06885$$

$Cos\beta Tan\lambda = 0.0647 < 0.1 = \mu$, hence it is confirmed that the condition for self locking is met.

Efficiency of the Worm

Lead angle of the worm is given by Equation [5.2.16]:

$$Tan\lambda = \frac{L}{\pi d_{pw}}$$

$$Tan\lambda = \frac{mn_w}{d_{pw}} = \frac{4}{58.1} = 0.0688 = 3°56'18''$$

Efficiency of the worm is given by Equation [[5.2.23]]

$$\eta = \frac{\left(Cos\beta - \mu Tan\lambda\right)}{Cos\beta\left(1 + \frac{r_b}{r_w}\mu_b Cot\lambda\right) + \mu Cot\lambda\left(1 - \frac{r_b}{r_w}\mu_b Tan\lambda\right)}$$

$$= \frac{Cos20 - 0.1.Tan3.938}{Cos20\left(1 + \frac{0.05}{0.029}Cot3.938\right) + 0.1.Cot3.938\left(1 - 0.02.\frac{0.05}{0.029}Tan3.938\right)} = 35.53\%$$

Hence torque in the worm shaft is:

$$T_w = \frac{1}{\eta_w}\frac{T_T}{GR} = \frac{1}{0.3553}\frac{1349.75}{72} = 52.76 \, Nm$$

Design of the Bevel Pair

Because the bevel gear and pinion are of the same material, the design of the bevel gear and pinion is based on the pinion dimensions. The design procedure follows that of spur gear design with some variation.

Number of teeth: It is recommended that pinion teeth should not be less than 17. Fixing the number of pinion teeth at this minimum value results in 85 gear teeth. Other combinations should be tried before the final selection is made.

Module range: An arbitrary range of standard modules from 1 to 10 is selected.

Pitch diameter: Pitch diameter is calculated for both bevel pinion and gear from Equation [5.2.12]. For a sample module 2:

For the pinion:

$$d_{pP} = \frac{mn}{10^3} = \frac{2.17}{1000} = 0.034 \text{ m}$$

For the gear:

$$d_{pG} = \frac{2.85}{1000} = 0.17 \text{ m}$$

Cone length: The cone length is calculated from Equation [6.2.1]:

$$l_c = \sqrt{\left(\frac{d_{pG}}{2}\right)^2 + \left(\frac{d_{pP}}{2}\right)^2} = \sqrt{\left(\frac{0.17}{2}\right)^2 + \left(\frac{0.034}{2}\right)^2} = 0.0867 \text{ m}$$

Tooth face width: Tooth face width is taken as $\frac{1}{3}l_c = 0.029$ m

Equivalent numbers of teeth: From Equation [6.2.5] for both pinion and gear:

$$n_E = \frac{2r}{mCos\theta} = \frac{T}{Cos\theta}$$

For pinion: $n_{EP} = \dfrac{T}{Cos\theta} = \dfrac{17}{Cos11°18'} = 17.337$

For gear: $n_{EG} = \dfrac{T}{Cos\theta} = \dfrac{85}{Cos78°41'} = 433.4$

Ratio Factor: Ratio Factor Q' is calculated from Equation [6.2.8]:

$$Q' = \frac{2n_{EG}}{n_{EP} + n_{EG}} = \frac{2.433.4}{433.4 + 17.337} = 1.92$$

Tangential force on pinion tooth: The tangential force on the pinion tooth is obtained by dividing the torque in the worm shaft by half the pitch diameter taken at the centre of the bevel gear tooth. From the already assumed relation of tooth width equal to one-third the cone length, the radius of the pitch circle at the centre of the tooth is $\frac{5}{6}r_{pG}$:

$$F_T = \frac{52.76}{0.0708} = 744.85 \text{ N}$$

Bending strength of the pinion tooth: Bending strength of the pinion tooth is got from Equation [6.2.6]:

The bending strength 103.4 M Pa of the steel is got from Fig 5.2.5 for steel of hardness 325 BHN. Since the operation is manual, the Velocity Factor K_v is one. The Lewis Form Factor Y' is interpolated from Fig 5.2.3 for the equivalent number of teeth.

$$F_S = oK_v bmY' \frac{l_c - b}{l_c} = 1146.63 \text{ N}$$

Limiting load for wear: This is obtained from Equation [6.2.7]

$$F_W = d'_p bK_w Q'$$

Here $d'_p = \frac{d_p}{Cos\theta}$ is the diameter of the rear cone circle, K_w the wear factor is obtained from Fig 5.2.5 and Q' the Ratio Factor is based on the equivalent number of pinion and gear teeth.

Selection of Module

The worksheet of Fig gives the following results for module 2:

Tangential tooth load: 762.21 N
Limiting bending load: 1146.63 N
Limiting load for wear: 2165.57 N

330

For module 2 the limiting loads are greater than the maximum tangential tooth load and the differences between the limiting loads and the tangential tooth load are minimal. Hence from strength and wear considerations module 2 offers the best option. However it is necessary to compute the torque at source before a final selection is made.

Worksheet for design of bevel gears: Selection of module							
torque Nm	53.99	GR	5	sigma MPa	103.4	Q	1.92
nP	17	nG	85	thetaP	0.197396		
Kv	1	BF	0.67	nEP	17.337		
Kw Mpa	1.124	Y	0.28784	nEG	433.417		
m mm	dpP m	dpG m	lc m	b m	FT N	FS N	FW N
1	0.017	0.085	0.043	0.014	1524.42	286.66	541.39
1.25	0.021	0.106	0.054	0.018	1219.54	447.90	845.93
1.5	0.026	0.128	0.065	0.022	1016.28	644.98	1218.14
2	0.034	0.170	0.087	0.029	762.21	1146.63	2165.57
2.5	0.043	0.213	0.108	0.036	609.77	1791.62	3383.71
3	0.051	0.255	0.130	0.043	508.14	2579.93	4872.54
4	0.068	0.340	0.173	0.058	381.11	4586.54	8662.29
5	0.085	0.425	0.217	0.072	304.88	7166.46	13534.83
6	0.102	0.510	0.260	0.087	254.07	10319.70	19490.16
8	0.136	0.680	0.347	0.116	190.55	18346.14	34649.17
10	0.170	0.850	0.433	0.144	152.44	28665.85	28152.45

Fig 6.3.2: Worksheet for selection of module for bevel gears for traversing gear train

Torque at the Source

Given the efficiency of the bevel pair as 0.98, the torque at the hand wheel is:

$$T_s = \frac{1}{\eta_b} \frac{T_w}{GR} = \frac{1}{0.98} \frac{52.76}{5} = 10.76 \text{ N-m}$$

The torque at source is almost double the acceptable hand wheel torque for comfortable manual operation. The best solution to reduce this torque is to increase the gear ratio between the bevel gear and pinion. The next logical gear ratio is 720, which will result in a traverse of one degree for two turns of the hand wheel. The worksheet of Fig 6.3.3 is recomputed for a gear ratio of 10. Calculations are given in the worksheet of Fig 6.3.4 below.

Worksheet for design of bevel gears: Selection of module							
torque Nm	52.76	GR	10	sigma MPa	103.4	Q	1.98
nP	17	nG	170	thetaP	0.099669		
Kv	1	BF	0.67	nEP	17.085		
Kw Mpa	1.124	Y'	0.28784	nEG	1708.479		
m mm	dpP m	dpG m	lc m	b m	FT N	FS N	FW N
1	0.017	0.170	0.085	0.028	744.85	564.99	1082.79
1.25	0.021	0.213	0.107	0.036	496.56	882.79	1691.85
1.5	0.026	0.255	0.128	0.043	413.80	1271.22	2436.27
2	0.034	0.340	0.171	0.057	310.35	2259.95	4331.15
2.5	0.043	0.425	0.214	0.071	248.28	3531.17	6767.42
3	0.051	0.510	0.256	0.085	206.90	5084.89	9745.08
4	0.068	0.680	0.342	0.114	155.18	9039.80	17324.59
5	0.085	0.850	0.427	0.142	124.14	14124.69	27069.67
6	0.102	1.020	0.513	0.171	103.45	20339.55	38980.32
8	0.136	1.360	0.683	0.228	77.59	36159.20	69298.35
10	0.170	1.700	0.854	0.285	62.07	56498.74	54680.73

Fig 6.3.3: Worksheet for selection of module for bevel gears with increased gear ratio.

The torque at source is in this case:

$$T_s = \frac{1}{0.98}\frac{52.76}{10} = 5.38 \text{ N-m, this is within the acceptable limit.}$$

The module selected here is 1.25 for which the loads are:

Tangential tooth load: 496.56 N

Limiting bending load: 882.79 N
Limiting load for wear: 1691.85 N

6.4 Design of a Power Traversing Gear Train

Design Exercise 6.4.1

This exercise is intended to illustrate the basic concepts involved in the preliminary design of a traversing gear train for heavy armament. It involves torque calculations, selection of gear parameters and definition of basic traversing motor characteristics. It is desired to design the traversing mechanism for a turret mounted gun on a vehicular chassis. The rate of fire of the gun is 10 rounds per minute and the gun is required to achieve the maximum traverse within the round to round interval. The maximum ground inclination, which the gun is expected to fire from, is 20°. The bore axis and the point of application of the recoil braking force both lie on the vertical plane through the traversing axis. The weight of the turret with all attachments is 89676.00 N. Effective radius of the ball race, of the turret ring is 0.8175 m. Coefficient of friction at the ball race is 0.003. Mass density of steel selected for the gears is 7700 Kg/m³. Distance from traversing axis to centre of gravity of the traversing parts is 0.127 m. Mass moment of inertia of the traversing parts is 3000.00 Kg-m². Acceleration of the traversing parts to be achieved is 0.5 rad/sec². The rated speed of the traversing motor is 6400 RPM.

Solution

The rate of fire being 10 rounds per minute, the round to round interval is 6 s. If in 6 s the gun is to be traversed by the maximum arc of 180°, it follows that the speed of traverse necessary is 30°/s.

Torque due to bearing friction as given by Equation [6.3.3] Reference 1 is:

$$T_b = \mu F r_b = 0.003 . 0.8175 . 89676.0 = 219.93 \ \text{Nm}$$

Weight moment due to ground inclination and angle of traverse vide Equation [6.3.1] Reference 1, is:

$$M_T = W_t r Sin\,\theta Cos\,\psi = 89676.0 . 0.0.127 . Sin20 . Cos0 = 3895.21 \ \text{N-m}$$

Torque necessary to accelerate the traversing parts given by Equation [6.3.4a] Reference 1 is:

$$T_{aT} = I_T \alpha_T = 3000.0.5 = 1500.00 \text{ N-m}$$

The net traversing torque is hence:

$$T_T = T_b + T_{aT} + M_T = 5614.14 \text{ N-m}$$

The necessary overall gear ratio is given by the ratio of the motor rated speed to the maximum traversing speed:

The traversing speed is 30º/s $= \dfrac{30}{360} 60 = 5$ RPM, so the overall gear ratio is:

$$GR = \frac{6400}{5} = 1280$$

The selection of the gear train configuration depends on space and physical displacement of the traversing ring from the power source and controls. An overall compact gear train is desirable with the least number of gear pairs. Minimizing the size of individual gears however indicates an increase in the number of meshing pairs. The gear train is designed to comprise of spur gear and pinion pairs with an efficiency of 0.98 at each mesh. Schematic of the gear train is depicted in Fig 6.4.1 ahead.

For the purpose of this exercise the breakdown of the overall gear ratio is selected as given below:

Traversing ring and Pinion No 1: 10
Gear No 2 and Pinion No 3: 4
Gear No 4 and Pinion No 5: 4
Gear No 6 and Pinion No 7: 8

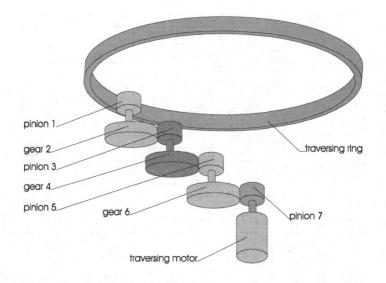

pinion 1

gear 2

pinion 3

gear 4

pinion 5

gear 6

traversing ring

pinion 7

traversing motor

Fig 6.4.1: Schematic of gear train for power traversing mechanism

Selection of Module

The selection of module for meshing pairs is done exactly as in the case of the elevating gear train in Design Exercise 5.4.1. The computations are given in the worksheets of Figs 6.4.2 to 6.4.5 ahead.

335

Worksheet for determining face width and number of teeth
Traversing ring and Pinion No 1

Gear ratio	10	Trav ring speed RPM	5	sigmap MPa			345
Min teeth on pinion	18	Torque Nm	5614.14	Speed Pinion No 1 RPM		50	
Min teeth on ring	180	Power tx W	2939.56				

Pinion teeth	Ring teeth	Pitch diameter m						Velocity m/s					
		2.5	3	4	5	6	8	2.5	3	4	5	6	8
18	180	0.045	0.054	0.072	0.090	0.108	0.144	0.12	0.14	0.19	0.24	0.28	0.38
24	240	0.060	0.072	0.096	0.120	0.144	0.192	0.16	0.19	0.25	0.31	0.38	0.50
30	300	0.075	0.090	0.120	0.150	0.180	0.240	0.20	0.24	0.31	0.39	0.47	0.63
36	360	0.090	0.108	0.144	0.180	0.216	0.288	0.24	0.28	0.38	0.47	0.57	0.75
42	420	0.105	0.126	0.168	0.210	0.252	0.336	0.27	0.33	0.44	0.55	0.66	0.88
48	480	0.120	0.144	0.192	0.240	0.288	0.384	0.31	0.38	0.50	0.63	0.75	1.01
54	540	0.135	0.162	0.216	0.270	0.324	0.432	0.35	0.42	0.57	0.71	0.85	1.13

Pinion teeth	Ring teeth	Velocity factor						Transmitted load KN					
		2.5	3	4	5	6	8	2.5	3	4	5	6	8
18	180	0.98	0.98	0.97	0.96	0.95	0.94	24.95	20.79	15.59	12.48	10.40	7.80
24	240	0.97	0.97	0.96	0.95	0.94	0.92	18.71	15.59	11.70	9.36	7.80	5.85
30	300	0.97	0.96	0.95	0.94	0.93	0.91	14.97	12.48	9.36	7.49	6.24	4.68
36	360	0.96	0.95	0.94	0.93	0.91	0.89	12.48	10.40	7.80	6.24	5.20	3.90
42	420	0.96	0.95	0.93	0.92	0.90	0.87	10.69	8.91	6.68	5.35	4.46	3.34
48	480	0.95	0.94	0.92	0.91	0.89	0.86	9.36	7.80	5.85	4.68	3.90	2.92
54	540	0.94	0.93	0.91	0.89	0.88	0.84	8.32	6.93	5.20	4.16	3.47	2.60

Pinion teeth	Ring teeth	Lewis Factor Y	Tooth width b mm					
			2.5	3	4	5	6	8
18	180	0.29327	100.58	70.12	39.74	25.63	17.93	10.24
24	240	0.33056	67.36	47.01	26.71	17.27	12.11	6.95
30	300	0.35510	50.48	35.28	20.09	13.02	9.15	5.27
36	360	0.37229	40.38	28.25	16.13	10.48	7.38	4.27
42	420	0.38703	33.50	23.47	13.43	8.74	6.17	3.59
48	480	0.39553	28.86	20.24	11.61	7.58	5.36	3.13
54	540	0.40335	25.32	17.78	10.22	6.68	4.74	2.77
Minimum face width			23.56	28.27	37.70	47.12	56.55	75.40
Maximum face width			39.27	47.12	62.83	78.54	94.25	125.66

Fig 6.4.2: Worksheet for selection of module for traversing ring and pinion 1

Standard modules selected for the analysis are 2.5, 3, 4, 5, 6, and 8. Based on criterion stated earlier, the module selected for the pair in question is 2.5, the tooth width towards the centre of the permissible range is 28.86 mm; number of pinion teeth is 48 and that of the traversing ring 480.

Worksheet for determining face width and number of teeth													
Gear 2 and pinion 3													
Gear ratio	4	Gear 2 speed RPM		50	sigmap MPa	345							
Min teeth on pinion	18	Torque Nm		572.87	Pinion speed RPM	200							
Min teeth on gear	72	Power tx W		2999.55									
Pinion	Gear	Pitch diameter m						Velocity m/s					
teeth	teeth	1.5	2	2.5	3	4	5	1.5	2	2.5	3	4	5
18	72	0.027	0.036	0.045	0.054	0.072	0.090	0.28	0.38	0.47	0.57	0.75	0.94
24	96	0.036	0.048	0.060	0.072	0.096	0.120	0.38	0.50	0.63	0.75	1.01	1.26
30	120	0.045	0.060	0.075	0.090	0.120	0.150	0.47	0.63	0.79	0.94	1.26	1.57
36	144	0.054	0.072	0.090	0.108	0.144	0.180	0.57	0.75	0.94	1.13	1.51	1.88
42	168	0.063	0.084	0.105	0.126	0.168	0.210	0.66	0.88	1.10	1.32	1.76	2.20
48	192	0.072	0.096	0.120	0.144	0.192	0.240	0.75	1.01	1.26	1.51	2.01	2.51
54	216	0.081	0.108	0.135	0.162	0.216	0.270	0.85	1.13	1.41	1.70	2.26	2.83

Pinion	Gear	Velocity factor						Transmitted load KN					
teeth	teeth	1.5	2	2.5	3	4	5	1.5	2	2.5	3	4	5
18	72	0.95	0.94	0.93	0.91	0.89	0.86	10.61	7.96	6.37	5.30	3.98	3.18
24	96	0.94	0.92	0.91	0.89	0.86	0.83	7.96	5.97	4.77	3.98	2.98	2.39
30	120	0.93	0.91	0.88	0.86	0.83	0.79	6.37	4.77	3.82	3.18	2.39	1.91
36	144	0.91	0.89	0.86	0.84	0.80	0.76	5.30	3.98	3.18	2.65	1.99	1.59
42	168	0.90	0.87	0.85	0.82	0.77	0.73	4.55	3.41	2.73	2.27	1.70	1.36
48	192	0.89	0.86	0.83	0.80	0.75	0.70	3.98	2.98	2.39	1.99	1.49	1.19
54	216	0.88	0.84	0.81	0.78	0.73	0.68	3.54	2.65	2.12	1.77	1.33	1.06

Pinion	Gear	LF	Tooth width b mm					
teeth	teeth	Y	1.5	2	2.5	3	4	5
18	72	0.29327	73.20	41.79	27.14	19.12	11.07	7.28
24	96	0.33056	49.43	28.35	18.50	13.09	7.64	5.06
30	120	0.35510	37.36	21.52	14.10	10.02	5.89	3.93
36	144	0.37229	30.13	17.43	11.47	8.18	4.84	3.26
42	168	0.38703	25.20	14.64	9.67	6.92	4.13	2.79
48	192	0.39553	21.88	12.76	8.46	6.08	3.65	2.48
54	216	0.40335	19.34	11.33	7.54	5.43	3.28	2.24
Minimum face width			14.14	18.85	23.56	28.27	37.70	47.12
Maximum face width			23.56	31.42	39.27	47.12	62.83	78.54

Fig 6.4.3: Worksheet for selection of module for gear No 2 and pinion No 3

Standard modules selected for the analysis are 1.5, 2, 2.5, 3, 4 and 5. The module selected for the pair in question is 1.5, the tooth width towards the centre of the permissible range is 21.88 mm, and number of pinion teeth is 48 and that of the gear 192.

Worksheet for determining face width and number of teeth						
Gear 4 and pinion 5						
Gear ratio	4	Gear 4 speed RPM	200	sigmap MPa		345
Min teeth on pinion	18	Torque	146.14	Pinion speed RPM		800
Min teeth on gear	72	Power tx	3060.76			

Pinion teeth	Gear teeth	Pitch diameter						Velocity					
		1	1.25	1.5	2	2.5	3	1	1.25	1.5	2	2.5	3
18	72	0.018	0.023	0.027	0.036	0.045	0.054	0.75	0.94	1.13	1.51	1.88	2.26
24	96	0.024	0.030	0.036	0.048	0.060	0.072	1.01	1.26	1.51	2.01	2.51	3.02
30	120	0.030	0.038	0.045	0.060	0.075	0.090	1.26	1.57	1.88	2.51	3.14	3.77
36	144	0.036	0.045	0.054	0.072	0.090	0.108	1.51	1.88	2.26	3.02	3.77	4.52
42	168	0.042	0.053	0.063	0.084	0.105	0.126	1.76	2.20	2.64	3.52	4.40	5.28
48	192	0.048	0.060	0.072	0.096	0.120	0.144	2.01	2.51	3.02	4.02	5.03	6.03
54	216	0.054	0.068	0.081	0.108	0.135	0.162	2.26	2.83	3.39	4.52	5.65	6.79

Pinion teeth	Gear teeth	Velocity factor						Transmitted load KN					
		1	1.25	1.5	2	2.5	3	1	1.25	1.5	2	2.5	3
18	72	0.89	0.86	0.84	0.80	0.76	0.73	4.06	3.25	2.71	2.03	1.62	1.35
24	96	0.86	0.83	0.80	0.75	0.70	0.67	3.04	2.44	2.03	1.52	1.22	1.01
30	120	0.83	0.79	0.76	0.70	0.66	0.61	2.44	1.95	1.62	1.22	0.97	0.81
36	144	0.80	0.76	0.73	0.67	0.61	0.57	2.03	1.62	1.35	1.01	0.81	0.68
42	168	0.77	0.73	0.69	0.63	0.58	0.53	1.74	1.39	1.16	0.87	0.70	0.58
48	192	0.75	0.70	0.67	0.60	0.54	0.50	1.52	1.22	1.01	0.76	0.61	0.51
54	216	0.73	0.68	0.64	0.57	0.51	0.47	1.35	1.08	0.90	0.68	0.54	0.45

Pinion teeth	Gear teeth	LF	Tooth width b					
		Y	1	1.25	1.5	2	2.5	3
18	72	0.29327	45.16	29.71	21.19	12.55	8.44	6.14
24	96	0.33056	31.17	20.66	14.85	8.91	6.06	4.46
30	120	0.35510	24.05	16.06	11.61	7.05	4.85	3.60
36	144	0.37229	19.77	13.29	9.67	5.94	4.12	3.08
42	168	0.38703	16.85	11.40	8.34	5.17	3.61	2.72
48	192	0.39553	14.89	10.13	7.45	4.66	3.28	2.49
54	216	0.40335	13.39	9.16	6.77	4.26	3.02	2.30
Minimum face width			9.42	11.78	14.14	18.85	23.56	28.27
Maximum face width			15.71	19.63	23.56	31.42	39.27	47.12

Fig 6.4.4: Worksheet for selection of module for gear No 4 and pinion No 5

Standard modules selected for the analysis are 1, 1.25, 1.5, 2, 2.5 and 3. The module selected for the pair in question is 1, the tooth width towards the centre of the permissible range is 13.39 mm, number of pinion teeth is 54 and that of the gear 216.

Worksheet for determining face width and number of teeth					
Gear 6 and pinion 7					
Gear ratio	8	Gear 6 speed RPM	800	sigmap MPa	345
Min teeth on pinion	18	Torque Nm	37.28	Pinion speed RPM	6400
Min teeth on gear	144	Power tx w	3123.23		

Pinion	Gear	Pitch diameter						Velocity					
teeth	teeth	1	1.25	1.5	2	2.5	3	1	1.25	1.5	2	2.5	3
18	144.00	0.018	0.023	0.027	0.036	0.045	0.054	6.03	7.54	9.05	12.06	15.08	18.10
24	192.00	0.024	0.030	0.036	0.048	0.060	0.072	8.04	10.05	12.06	16.08	20.11	24.13
30	240.00	0.030	0.038	0.045	0.060	0.075	0.090	10.05	12.57	15.08	20.11	25.13	30.16
36	288.00	0.036	0.045	0.054	0.072	0.090	0.108	12.06	15.08	18.10	24.13	30.16	36.19
42	336.00	0.042	0.053	0.063	0.084	0.105	0.126	14.07	17.59	21.11	28.15	35.19	42.22
48	384.00	0.048	0.060	0.072	0.096	0.120	0.144	16.08	20.11	24.13	32.17	40.21	48.25
54	432.00	0.054	0.068	0.081	0.108	0.135	0.162	18.10	22.62	27.14	36.19	45.24	54.29

Pinion	Gear	Velocity factor						Transmitted load KN					
teeth	teeth	1	1.25	1.5	2	2.5	3	1	1.25	1.5	2	2.5	3
18	144.00	0.50	0.44	0.40	0.33	0.28	0.25	0.52	0.41	0.35	0.26	0.21	0.17
24	192.00	0.43	0.37	0.33	0.27	0.23	0.20	0.39	0.31	0.26	0.19	0.16	0.13
30	240.00	0.37	0.32	0.28	0.23	0.19	0.17	0.31	0.25	0.21	0.16	0.12	0.10
36	288.00	0.33	0.28	0.25	0.20	0.17	0.14	0.26	0.21	0.17	0.13	0.10	0.09
42	336.00	0.30	0.25	0.22	0.18	0.15	0.12	0.22	0.18	0.15	0.11	0.09	0.07
48	384.00	0.27	0.23	0.20	0.16	0.13	0.11	0.19	0.16	0.13	0.10	0.08	0.06
54	432.00	0.25	0.21	0.18	0.14	0.12	0.10	0.17	0.14	0.12	0.09	0.07	0.06

Pinion	Gear	LF	Tooth width b				
teeth	teeth	Y	1	1.25	1.5	2	2.5	3
18	144.00	0.29327	10.26	7.39	5.70	3.85	2.88	2.28
24	192.00	0.33056	7.97	5.83	4.56	3.13	2.37	1.90
30	240.00	0.35510	6.78	5.02	3.96	2.76	2.11	1.70
36	288.00	0.37229	6.07	4.53	3.60	2.53	1.94	1.57
42	336.00	0.38703	5.56	4.18	3.34	2.36	1.83	1.48
48	384.00	0.39553	5.24	3.96	3.18	2.26	1.75	1.43
54	432.00	0.40335	4.98	3.79	3.05	2.18	1.69	1.38
Minimum face width			9.42	11.78	14.14	18.85	23.56	28.27
Maximum face width			15.71	19.63	23.56	31.42	39.27	47.12

Fig 6.4.5: Worksheet for selection of module for gear No 6 and pinion No 7

Standard modules selected for the analysis are 1.5, 2, 2.5 and 3. The module selected for the pair in question is 1, the tooth width towards the centre of the permissible range is 10.26 mm, and the number of pinion teeth is 18 and that of the gear 144.

Torque at Source

The torque at the motor shaft is given by Equation [6.2.8] Reference 1:

$$T_s = \frac{1}{\eta^n}\frac{T_T}{GR} = \frac{1}{0.98^4}\frac{5614.14}{1280} = 4.755 \text{ N-m.}$$

The torque at source is within the acceptable limits for hand wheel operation in the case of power failure.

Torque Necessary to Accelerate the Traversing Gear Train

The torque necessary to accelerate the gear train itself is calculated for the selected modules and corresponding gear radii as for traversing gears in Design Exercise 5.4.1. The calculations are given in the worksheet of Fig 6.4.6 below.

Worksheet to compute torque for accelerating the gear train											
Mass density of steel Kg/m^3			7700.00	neta		0.98	alphaE	0.1			
Gear/ pinion	Teeth	module	b m	rp m	m Kg	IG	Shaft	I	alpha	T Nm	T source N m
1	48	2.5	0.029	0.0600	2.513	4.52E-03					
2	192	1.5	0.022	0.1440	10.975	1.14E-01	1 to 2	0.118315	1	0.1183	16.09056
3	48	1.5	0.022	0.0360	0.686	4.44E-04					
4	216	1.0	0.013	0.1080	3.778	2.20E-02	3 to 4	0.022478	4	0.0899	2.99584
5	54	1.0	0.013	0.0270	0.236	8.61E-05					
6	144	1.0	0.010	0.0720	1.287	3.33E-03	5 to 6	0.003421	16	0.0547	0.44683
7	18	1.0	0.010	0.0090	0.020	8.14E-07	7	0.000001	128	0.0001	0.00010
Torque necessary to accelerate the gear train Nm											19.53332
Torque necessary to accelerate the gear train as a % of torque at source Nm											4.10778

Fig 6.4.6: Worksheet for calculation of torque required to accelerate the traversing gear train

The torque is around 4 times the source torque hence is to be included in the torque calculations.

Required Motor Horsepower

Net torque required at the source is 4.755 + 19.53 Nm

The required power of the traversing motor is given by:

$$W_S = \frac{2\pi N T_S}{60000} = 16.28 \text{ KW or 22 HP}$$

APPENDICES

Useful Conversion Tables

Length

	m	in	ft	yd
m	1	39.37	3.281	1.094
in	$2.540.10^{-2}$	1	$8.333.10^{-2}$	$2.778.10^2$
Ft	$3.048.10^{-1}$	12	1	$3.333.10^{-1}$
Yd	$9.144.10^{-1}$	36	3	1

Area

	m^2	In^2	ft^2	yd^2
m^2	1	$1.550.10^3$	$1.076.10^1$	1.196
in^2	$6.452.10^{-4}$	1	$6.944.10^{-3}$	$7.716.10^{-4}$
ft^2	$9.290.10^{-2}$	$1.440.10^2$	1	$1.111.10^{-4}$
yd^2	$8.361.10^{-1}$	$1.296.10^3$	9	1

Mass

	kg	lb	ton(UK)	ton(US)	grain
kg	1	2.205	$9.842.10^{-4}$	$1.102.10^{-3}$	$1.543.10^4$
lb	$4.535.10^{-1}$	1	$4.464.10^{-4}$	$5.0.10^{-4}$	$7.0.10^3$
ton (UK)	$1.016.10^3$	$2.240.10^3$	1	1.120	$1.567.10^7$
ton (US)	$9.072.10^2$	$2.0.10^3$	$8.929.10^{-1}$	1	$1.40.10^7$
grain	$6.480.10^{-5}$	$1.428.10^{-4}$	$6.378.10^{-8}$	$7.141.10^{-8}$	1

Volume

	m^3	litre	in^3	ft^3	yd^3
m	1	10^3	$6.102.10^4$	35.31	1.308
litre	10^{-3}	1	61.02	$3.531.10^{-2}$	$1.308.10^{-3}$
in^3	$1.639\text{-}10^{-5}$	$1.639.10^{-2}$	1	$5.787.10^4$	$2.143.10^{-5}$
ft^3	$2.832.10^{-2}$	28.32	$1.728.10^3$	1	$3.704.10^{-2}$
yd^3	$7.646.10^{-1}$	$7.646.10^2$	$4.666.10^4$	27	1

Angular Velocity

	rad/s	rad/min	degree/s	rps	rpm
rad/s	1	60	57.30	$1.592.10^{-1}$	9.549
rad/min	$1.667.10^{-2}$	1	$9.550.10^{-1}$	$2.653.10^{-3}$	$1.592.10^{-1}$
degree/s	$1.745.10^{-2}$	1.047	1	$2.778.10^{-3}$	$1.667.10^{-1}$
rps	6.283	$3.770.10^{2}$	$3.60.10^{2}$	1	60
rpm	$1.047.10^{-1}$	6.283	60	$1.667.10^{-2}$	1

Velocity

	m/s	m/min	km/h	Ft/s	Ft/min	mile/h
m/s	1	60	3.6	3.281	$1.969.10^{2}$	2.237
m/min	$1.667.10^{-2}$	1	6.10^{-2}	$5.468.10^{-2}$	3.281	$3.728.10^{-2}$
km/h	$2.778.10^{-1}$	16.67	1	$9.113.10^{-1}$	54.68	$6.214.10^{-1}$
ft/s	$3.048.10^{-1}$	18.29	1.097	1	60	$6.818.10^{-1}$
ft/min	$5.080.10^{-3}$	$3.048.10^{-1}$	$1.829.10^{-2}$	$1.667.10^{-2}$	1	$1.136.10^{-2}$
mile/h	$4.470.10^{-1}$	$2.682.10^{1}$	1.609	1.467	$8.80.10^{4}$	1

Density

	kg/m³	ib/ft³	lb/in³
kg/m3	1	$6.243.10^{-2}$	$3.613.10^{-5}$
lb/ft³	16.02	1	$5.787.10^{-4}$
lb/in³	$2.768.10^{4}$	$1.728.10^{3}$	1

Acceleration

	m/s²	ft/s²
m/s²	1	3.281
ft/s²	$3.048.10^{-1}$	1

Force

	N	kgf	dyne	lbf	tonf UK	tonf US	poundal
N	1	$1.020.10^{-1}$	$1.0.10^{5}$	$2.248.10^{-1}$	$1.004.10^{-4}$	$1.124.10^{-4}$	7.233
kgf	9.807	1	$9.807.10^{5}$	2.205	$9.842.10^{-4}$	$1.102.10^{-3}$	70.93
dyne	1.10^{-5}	$1.020.10^{-6}$	1	$2.248.10^{-6}$	$1.004.10^{-9}$	$1.124.10^{-9}$	$7.233.10^{-5}$
lbf	4.448	$4.536.10^{-1}$	$4.448.10^{5}$	1	$4.464.10^{-4}$	5.10^{-4}	32.17
tonf UK	$9.964.10^{3}$	$1.016.10^{3}$	$9.964.10^{8}$	$2.240.10^{3}$	1	1.120	$7.207.10^{4}$
tonf US	$8.896.10^{3}$	$9.075.10^{2}$	$8.893.10^{3}$	2.10^{3}	$8.932.10^{-1}$	1	$6.433.10^{4}$
pdl	$1.383.10^{-1}$	$1.410.10^{-2}$	$1.383.10^{4}$	$3.108.10^{2}$	$1.388.10^{-5}$	$1.554.10^{-5}$	1

Moment of Force

	Nm	kgfm,	lbf ft
Nm	1	1.020^{-1}	$7.376.10^{-1}$
kgfm	9.807	1	7.233
lbf ft	1.356	$1.382.10^{-1}$	1

Energy

	J	kwh	ft lbf
J	1	$2.778.10^{-7}$	$7.376.10^{-1}$
kwh	$3.6.10^{6}$	1	$2.655.10^{6}$
ft lbf	1.356	$3.766.10^{-7}$	1

Pressure

	Pa	bar	Kgf/cm^2	atm	psi
Pa	1	1.10^{-5}	$1.020.10^{-5}$	$9.869.10^{-6}$	$1.450.10^{-4}$
bar	1.10^5	1	1.020	$9.869.10^{-1}$	14.51
kgf/cm^2	$9.807.10^4$	$9.807.10^{-1}$	1	$9.678.10^{-1}$	14.22
atm	$1.013.10^5$	1.013	1.033	1	14.70
psi	$6.895.10^3$	$6.895.10^{-2}$	$7.031.10^{-2}$	$6.805.10^{-2}$	1

Note: Multiples of Pa: GPa, MPa, KPa

Stress

	Pa	N/mm^2	N/cm^2	kgf/mm^2	kgf/mm^2	lbf/in^2	tonf/in^2 (UK)
Pa	1	1.10^{-6}	1.10^{-4}	$1.020.10^{-7}$	$1.020.10^{-5}$	$1.450.10^{-4}$	$6.475.10^{-8}$
N/mm^2	1.10^6	1	100	$1.020.10^{-1}$	10.20	145.00	$6.475.10^{-2}$
N/cm^2	1.10^4	1.10^{-2}	1	$1.020.10^{-3}$	$1.020.10^{-1}$	1.450	$6.475.10^{-4}$
kg/mm^2	$9.807.10^6$	9.807	$9.807.10^2$	1	100	1422.00	$6.350.10^{-1}$
kg/cm^2	$9.807.10^4$	$9.807.10^{-2}$	9.807	1.10^{-2}	1	14.22	$6.350.10^{-3}$
lbf/in^2	$6.895.10^3$	$6.895.10^{-3}$	$6.895.10^{-1}$	$7.031.10^{-4}$	$7.031.10^{-2}$	1	$4.464.10^{-4}$
tonf/in^2 (UK)	$1.544.10^7$	1.5.44	$1.544.10^3$	1.575	157.50	2240.00	1

Power

	w	HP	ft lbf/s
w	1	$1.341.10^{-3}$	$7.376.10^{-1}$
HP	$7.457.10^2$	1	$5.50.10^2$
ft lbf/s	1.356	$1.818.10^{-3}$	1

Moments of Inertia of Sections about the Neutral Axis as Shown

Description of Section	Moment of Inertia about the neutral axis
Rectangular lamina of breadth b and depth d	$\dfrac{bd^3}{12}$
Parallelogram of side a	$\dfrac{a^4}{12}$

Circular disc of diameter d	$$\frac{\pi}{64}d^4$$
Annular disc of outer diameter D and inner diameter d	$$\frac{\pi}{64}\left(D^4 - d^4\right)$$
Elliptical disc	$$\frac{\pi}{4}ab^3$$

Triangle of base b and height h 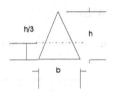	$\dfrac{1}{36}bh^3$

Moments of Inertia of Solids

Moments of inertia of various solid bodies of uniform density and of mass m.

Description of Body	Moment of inertia
Thin rod of length l • About the axis perpendicular to the rod through the center of mass, • About the axis perpendicular to the rod through one end.	$\frac{1}{12}ml^2$ $\frac{1}{3}ml^2$
Rectangular lamina with sides a, b • About an axis perpendicular to the plate through its center. • About an axis parallel to side b through its center.	$\frac{1}{12}m(a^2+b^2)$ $\frac{1}{12}ma^2$

Circular cylinder of radius r and length l • About axis of cylinder, • About axis through center of mass and perpendicular to cylindrical axis, • About axis coinciding with diameter at one end.	$\dfrac{1}{2}mr^2$ $\dfrac{1}{12}m(l^2 + 3r^2)$ $\dfrac{1}{12}m(4l^2 + 3r^2)$
Hollow circular cylinder of outer radius r_o, inner radius r_i and length l • About the axis of cylinder. • About the axis through center of mass and perpendicular to cylindrical axis. • About the axis coinciding with a diameter at one end.	$\dfrac{1}{2}m(r_0^2 + r_i^2)$ $\dfrac{1}{12}m(3r_i^2 + 3r_0^2 + l^2)$ $\dfrac{1}{12}m(3r_i^2 + 3r_0^2 + 4l^2)$

Circular plate of radius r • About axis perpendicular to plate through center. • About axis coinciding with a diameter.	$\dfrac{1}{2}mr^2$ $\dfrac{1}{4}mr^2$
Hollow circular plate or ring with outer radius r_0 and inner radius r_i • About axis perpendicular to plane of plate through center, • About axis coinciding with a diameter.	$\dfrac{1}{2}m(r_0^2 + r_i^2)$ $\dfrac{1}{4}m(r_0^2 + r_i^2)$
Thin circular ring of radius r • About axis perpendicular to plane of ring through center. • About axis coinciding with diameter.	mr^2 $\dfrac{1}{2}mr^2$
Sphere of radius r • About axis coinciding with a diameter, • About axis tangent to the surface.	$\dfrac{2}{5}mr^2$ $\dfrac{7}{5}mr^2$

Hollow sphere of outer radius r_o and inner radius r_i • About axis coinciding with a diameter. • About axis tangent to the surface.	$\dfrac{2}{5}m\left(r_o^5 - r_i^5\right)\left(r_o^3 - r_i^3\right)$ $\dfrac{2}{5}m\left(\dfrac{r_o^5 - r_i^5}{r_o^3 - r_i^3}\right) + mr_o^2$
Hollow spherical shell of radius r • about axis coinciding with a diameter, • About axis tangent to the surface.	$\dfrac{2}{3}mr^2$ $\dfrac{5}{3}mr^2$

Geometric Formulae

Triangle of altitude h and base b Area $= bh = ab\sin\theta = \sqrt{(s(s-a)(s-b)-(s-c))}$ Where: $s = \dfrac{1}{2}(a+b+c) =$ semiperimeter	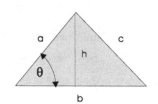
Sector of circle of radius r Area $= \dfrac{1}{2}r^2\theta$, θ is in radians Arc length $s = r\theta$	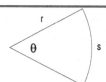
Segment of circle of radius r Area $= \dfrac{1}{2}r^2(\theta - \sin\theta)$	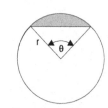

Ellipse of semi-major axis a and semi-minor axis b Area $= \pi ab$ Perimeter $=$ $4a \int_0^{\pi/2} \sqrt{1 - k^2 \sin^2 \theta \, d\theta} = 2\pi \sqrt{\frac{1}{2}(a^2 + b^2)}$ Where $k = \sqrt{a^2 - b^2}/a$.	
Segment of a parabola ABC Area $= \dfrac{2}{3} ab$ Arc length ABC $=$ $\dfrac{1}{2}\sqrt{b + 16a^2} + \dfrac{b^2}{8a} \ln\left(\dfrac{4a + \sqrt{b^2 + 16a^2}}{b} \right)$	
Right circular cone of radius r and height h Volume $= \dfrac{1}{3}\pi r^2 h$ Lateral surface area $= \pi r \sqrt{r^2 + h^2} = \pi r l$	
Spherical cap of radius r and height h Volume (shaded) $= \dfrac{1}{3}\pi h^2 (3r - h)$ Surface area $= 2\pi r h$	

Frustum of right circular cone of radii r_1, r_2 and height h Volume $= \dfrac{1}{3}\pi h(r_1^2 + r_1 r_2 + r_2^2)$ Lateral surface area $= \pi(r_1 + r_2)\sqrt{h^2 + (r_2 - r_1)^2}$ $= \pi(r_1 + r_2)l$	
Spherical triangle of angles A, B, C on sphere of radius r Area of triangle $ABC = (A + B + C - \pi)r^2$	
Torus of inner radius r_i and outer radius r_o Volume $= \dfrac{1}{4}\pi^2(r_o + r_i)(r_o - r_i)^2$ Surface area $= \pi^2(r_o^2 - r_i^2)$	
Ellipsoid of semi-axes a,b,c Volume $= \dfrac{4}{3}\pi abc$	

Paraboloid of revolution

Volume $= \dfrac{1}{2}\pi b^2 a$

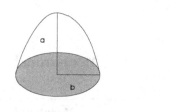

Mechanical Properties of Common Metals

Metal	Modulus of Elasticity E GPa	Modulus of Rigidity G GPa	Poisson's Ratio μ	Density ρ Kg/m³
Aluminum alloy	72	27	0.32	2800
Beryl Copper	127	50	0.29	8300
Brass/bronze	110	41	0.33	8700
Copper	121	46	0.33	8900
Cast iron	103	41	0.26	7200
Magnesium alloy	45	17	0.35	1800
Nickel alloy	207	79	0.3	8300
Carbon steel	207	79	0.3	7700
Steel alloy	207	79	0.3	7700
Stainless steel	190	73	0.3	7700
Titanium alloy	114	43	0.33	4400
Zinc alloy	83	31	0.33	6600

Note: Data is approximate, values will differ with composition & processing

Mechanical Properties of Popular Carbon & Alloy Steels

AISI No	Treatment	Tensile strength MPa	Yield Strength MPa	Elongation %	Reduction in Area %	BHN	Izod Impact Strength J
1015	Rolled	420.6	313.7	39.0	61.0	126	110.5
	Normalized	424.0	324.1	37.0	69.6	121	115.5
	Annealed	386.1	284.4	37.0	69.7	111	115.0
1020	Rolled	448.2	330.9	36.0	59.0	143	86.8
	Normalized	441.3	346.5	35.8	67.9	131	117.7
	Annealed	394.7	294.8	36.5	66.0	111	123.4
1030	Rolled	551.6	344.7	32.0	57.0	179	74.6
	Normalized	520.6	344.7	32.0	60.8	149	93.6
	Annealed	463.7	341.3	31.2	57.9	126	69.4
1040	Rolled	620.5	413.7	25.0	50.0	201	48.8
	Normalized	589.5	374.0	28.0	54.9	170	65.1
	Annealed	518.8	353.4	30.2	57.2	149	44.3
1050	Rolled	723.9	413.7	20.0	40.0	229	31.2
	Normalized	748.1	427.5	20.0	39.4	217	27.1
	Annealed	636.0	365.4	23.7	39.9	187	16.9
1095	Rolled	965.3	572.3	9.0	18.0	293	4.1
	Normalized	1013.5	499.9	9.5	13.5	293	5.4
	Annealed	656.7	379.2	13.0	20.6	192	2.7
1118	Rolled	521.2	316.5	32.0	70.0	149	108.5
	Normalized	477.8	319.2	33.5	65.9	143	103.4
	Annealed	450.2	284.8	34.5	66.8	131	106.4

Mechanical Properties of Popular Carbon & Alloy Steels
(continued)

AISI No	Treatment	Tensile strength MPa	Yield Strength MPa	Elongation %	Reduction in Area %	BHN	Izod Impact Strength J
3140	Normalized	891.5	599.8	19.7	57.3	262	53.6
	Annealed	689.5	422.6	24.5	50.8	197	46.4
4130	Normalized	668.8	436.4	25.5	59.5	197	86.4
	Annealed	560.5	360.6	28.2	55.6	156	61.7
4140	Normalized	1020.4	655.0	17.7	46.8	302	22.6
	Annealed	655.0	417.1	25.7	56.9	197	54.5
4340	Normalized	1279.0	861.8	12.2	36.3	363	15.9
	Annealed	744.6	472.3	22.0	49.9	217	51.1
6150	Normalized	939.8	615.7	21.8	61.0	269	35.5
	Annealed	667.4	412.3	23.0	48.4	197	27.4
8650	Normalized	1023.9	688.1	14.0	40.4	302	13.6
	Annealed	715.7	386.1	22.5	46.4	212	29.4
8740	Normalized	929.4	606.7	16.0	47.9	269	17.6
	Annealed	695.0	415.8	22.2	46.4	201	40.0
9255	Normalized	932.9	579.2	19.7	43.4	269	13.6
	Annealed	774.3	486.1	21.7	41.1	229	8.8

Note: Data refers to mean values of 1" i.e. 25.4 mm round sections

G

References
(In alphabetical order)

Armament Engineering a computer aided approach, H Peter, Trafford, 2003
122mm Howitzer D 30, Service Manual
130mm gun Model 1946, Service Manual
76 mm Gun Model 1942, Service Manual
Advanced Fluid Mechanics, Volume 1, 1958, RC Binder, Prentice Hall Inc.
Applied Fluid Mechanics, 2nd Edition 1979, Robert L Mott, Charles E Merill Publishing Company
Elements of Strength of Materials, S Timoshenko & Gleason H MacCullough, 3rd Edition 1949, D Van Nortrand Company Inc.
Foundations of Fluid Mechanics, SW Yuan, Prentice Hall of India, 1988.
Handbook of Metals, American Society for Metals, 1985
Introduction to Theory of Plasticity for Engineers, O Hoffman & G Sachs, Mc Graw-Hill Book Company 1953
Machine Design An Integrated Approach, 1996, Robert L Norton, Prentice Hall
Machine Design, Paul H Black, O Eugene Adams Jr, McGraw-Hill 1981
Matlab for Engineering Applications, William J Paul III, WCB/McGraw-Hill 1999
Mechanics of Engineering Materials, 2nd Edition 1997, PP Benham, RJ Crawford, CG Armstrong, Addison Wesley Longman Ltd.
Mechanics of Materials 3rd Edition 1997, EJ Hearn, Butterworth Heinmann
Microsoft Excel in 24 Hours, Trudi Reisner, Techmedia
Schaums Outline Series Theory and Problems of Strength of Materials 2nd Edition, 1972, William A Nash, McGraw Hill Inc.
Solving Problems in Fluid Mechanics, JF Douglas, Longman Scientific & Technical, 1989.
Strength of Materials, 1973, V Feodosyev, Mir Publishers
Tanks Design & Calculation, SS Bunov, Mir Publications
Text Book of Service Ordnance 1923
The Student Edition of Matlab Version 4, Users Guide, The Mathworks Inc, Prentice Hall Inc
Theory of Elasticity, SP Timoshenko & JN Goodier, 3rd Edition 1970, McGraw Hill
Theory of Plates & sShells, SP Timoshenko, S Woinowsky-Krieger, McGraw-Hill, 1959

INDEX

A

absolute gas pressure 248,258,260
accuracy 2, 3, 4, 5, 68, 73, 76, 91, 92, 100, 107, 275, 277
actual gas pressure 248,249,252,258,260,342
addendum 286, 289
after effect 148
allowable stress 14
ammunition 5, 79, 80, 81, 82, 86, 108, 109, 110, 120
angle of rifling 65, 66, 68, 73
area-moment method 92
autofrettage
swage 34, 63
autofrettaged 32, 37, 50, 51, 55
axial stress 3, 17, 47, 83

B

balancing gear 221, 222, 223, 224, 227, 228, 229, 230, 233, 241, 242, 243, 244, 245, 246, 249, 250, 253, 254, 255, 256, 257, 258, 259, 260, 263, 264, 266, 267, 268, 272, 273, 274, 278, 279, 312
barrel
  bend 5
  cooling 5
  dilation 4
  nut 132
  wall thickness 3, 15
  whip 3
Barth Equation 285
Barth's Velocity Factor 285, 303
bending moment 93, 94, 95, 101
bore 2, 3, 4, 14, 15, 17, 20, 22, 32, 35, 36, 43, 51, 65, 66, 73, 75, 78, 79, 80, 84, 87, 115, 143, 166, 222, 224, 250, 275, 276, 311, 317, 319, 323, 333
bottom carriage 222, 241, 242, 243, 245, 252, 254, 255, 263, 266, 267, 312, 313
braking force 148, 152, 153, 154, 155, 157, 161, 162, 171, 172, 174, 176, 178, 179, 180, 186, 189, 191, 279, 317, 333
breech 3, 14, 16, 22, 30, 80, 85, 108, 109, 110, 111, 112, 113, 114, 115, 116, 117, 118, 119, 120, 121, 126, 127, 128, 129, 130, 131,

132, 133, 136, 137, 141, 142, 143, 146, 147, 148
  assembly 108, 109, 110, 132
  ring 16, 108, 109, 113, 115, 116, 117, 132
  screw 109, 110, 113, 114, 115, 116, 117, 121, 128, 131
buffer 164, 186, 189, 194, 195, 198, 199, 200, 201, 202, 203, 204, 209, 210, 211
  stroke 189, 194, 195, 198, 200, 204
buffing phase 197, 204

C

calibre 1, 19, 20, 68, 74, 76, 78, 79, 80, 81, 87, 88, 109, 121, 152
cam 111, 134, 136
cantilever 93, 95, 98, 282
cartridge case 80, 81, 83, 84, 85, 86, 87, 109, 110, 111, 120, 121, 132, 133, 142, 143
centering cylinder 80
chamber 4, 5, 13, 14, 19, 22, 29, 31, 32, 35, 51, 79, 80, 81, 82, 83, 84, 85, 86, 87, 88, 89, 90, 109, 110, 111, 114, 120, 121, 132, 148
  length 79, 87, 88
  mouth 79, 80, 87
  face 80
  slope 80
  volume 80
chamfer 74,79
charge mass 20, 69, 154, 166
chromium 2
circular frequency 99
circular pitch 281
circumferential strain 40,41,123
clearance 4, 35, 36, 39, 61, 65, 80, 82, 84, 85, 87, 88, 109, 111, 131, 143, 152
closing spring 113, 117, 118, 120, 121, 133, 134, 141, 142, 143
cluster gear 315
clutch 277, 278, 315
CMP 13, 14, 20, 21, 22, 25
cocking 109, 113, 118, 119, 120, 132, 143
combined twist 65
conduction 4
cone centre 319
cone length 329, 330
constant orifice 198
container 35, 37, 39, 41, 61, 62

369

continuous pull 118
conventions 13
cook off 5
counter recoil 109, 111, 112, 119, 132, 133, 135, 136, 137, 143, 145, 148, 149, 150, 152, 164, 182, 183, 184, 185, 186, 187, 188, 189, 190, 191, 192, 194, 195, 196, 197, 198, 200, 201, 202, 203, 204, 205, 206, 208, 209, 210, 211, 212
cradle 132, 147, 149, 197, 222, 241, 242, 243, 244, 246, 252, 254, 255, 263, 266, 267, 278, 312
crank
roller 112, 113, 132
  shaft 111, 113, 115, 119, 132, 133, 134, 135, 136, 137, 138, 139, 140, 141, 142, 143, 144, 145, 146
cross head 115
cylinder 17, 35, 50, 80, 81, 83, 117, 150, 151, 162, 178, 188, 198, 200, 213, 215, 216, 217, 218, 219, 220, 241, 281

D

dedendum 286, 289
deflection 41, 42, 92, 93, 94, 95, 122, 127, 222, 224, 225
diametral pitch 281
Direct Shear Factor 265, 270
Discharge Coefficient 164, 204
driving band 2, 4, 64, 65, 68, 72, 73, 74, 78, 79, 80, 81, 91
driving edge 65, 78
droop 3, 16

E

eccentricity 15, 38, 317
efficiency 290, 300, 328
ejection 109
elastic limit 32,35,47,81,82
elevating mechanisms 275, 276, 316
engraving 73, 75
epicyclic gear 315
equivalent number of teeth 321, 322, 330
erosion 2
ESP 14, 16, 26, 27, 28, 37, 38, 52, 54
extraction 79, 80, 82, 83, 84, 85, 109

extractor 80, 120, 121, 132, 133, 142

F

face width 282, 283, 288, 290, 297, 303, 304, 305, 306, 307, 308, 309, 322, 325, 326, 329
firing mechanism 108, 109, 110, 113, 118, 119, 132
firing pin 118, 119, 120
flange 80, 85
follower 111, 113, 132, 133, 134, 135, 136, 137, 138, 142, 143, 145
forcing cone 73, 79, 80, 81, 82, 87, 91
Frequency Ratio 91, 92, 107
friction 17, 73, 85, 114, 118, 148, 152, 153, 157, 160, 161, 162, 168, 176, 177, 178, 182, 183, 188, 191, 192, 193, 194, 195, 196, 204, 207, 222, 240, 241, 242, 243, 244, 245, 246, 248, 249, 251, 252, 253, 254, 255, 256, 257, 258, 259, 261, 262, 263, 266, 267, 276, 277, 279, 281, 282, 292, 293, 294, 295, 298, 300, 316, 323, 333
torque 266, 267

G

gas pressure 2, 3, 17, 20, 34, 51, 55, 57, 58, 61, 63, 69, 80, 81, 83, 85, 114, 125, 126, 128, 129, 131, 132, 148, 166, 177, 241, 242, 248, 249, 258, 260, 295, 298
gear
bevel 320
ratio 276, 277, 280, 295, 296, 297, 301, 302, 324, 332, 334
spur 286, 288, 300, 314, 321, 322, 328, 334
worm 289, 290, 291, 292, 294, 295, 296, 297, 298, 299, 325
train 276, 295, 296, 300, 308, 333, 340
gears
balancing 221, 237, 263, 278, 279, 281, 289, 319, 320, 321
groove 65, 73, 74, 75, 76, 78, 79, 81, 164
groove root 65
groove width 74
gun barrel 1, 6, 13, 16, 17, 19, 29, 32, 51, 55, 63, 86, 92, 93, 95, 98, 100

monobloc 14, 17, 20, 22, 26, 28, 30, 32, 38, 51, 52, 60, 61
gun pressure codes 13

H

hammer 118, 119
harmonic motion 96
hoop stress 17, 33, 34, 39, 41, 42, 55, 56, 84, 123, 128, 130, 216, 217, 220
Huber-Von Mises-Hencky 6, 7, 12, 18
hydraulic 34, 35, 36, 43, 61, 148, 150, 151, 152, 153, 161, 162, 176, 178, 179, 180, 181, 184, 186, 189, 191, 213, 222, 276, 278

I

impact 81, 116, 118, 120, 132, 148, 197, 204
impact velocity 81
increasing twist 68,73,
inertia percussion 118,
interface32, 33, 40, 41, 42, 48, 49, 55, 62, 160, 162
interference 42, 50, 62, 64, 82
internal ballistics 13, 20, 22, 69, 79, 155, 166, 168
interrupted thread 114, 121

L

land 19, 73, 74, 75, 76, 78, 286
width 74, 75
Lewis Equation 282, 321
Lewis Form Factor 283, 284, 285, 303, 325, 330
locking 109, 113, 114, 133, 290, 294, 297, 324, 327
lump loading 98

M

mandrel 34, 43, 50, 63, 64
manual depression 249, 255, 256, 261
manual elevation 223, 239, 247, 248, 255, 257, 259
mass density 100, 333
mating 16, 113, 114, 319

maximum shear stress 6, 7, 33, 39, 216, 264
method of least squares 232
mid section 29, 30, 31
module 281, 297, 303, 322, 324, 326, 328, 330, 335
modulus of rigidity 46, 63
moment of inertia 93, 101, 280, 300, 309, 318, 333
muzzle 1, 3, 5, 13, 16, 20, 29, 30, 31, 64, 65, 69, 73, 91, 147, 148, 154, 155, 165, 166, 227, 228, 238
    preponderence 227,228,231,238,
    reference 5
    velocity 13, 20, 69, 154, 165, 166

N

natural frequency 91, 92, 97, 98, 99, 106, 107
nickel 2

O

open-end effects 91
operating mechanism 108, 109, 110, 111, 112, 115, 116, 133
orifice 152, 162, 163, 164, 165, 166, 168, 180, 181, 182, 184, 186, 191, 192, 194, 195, 196, 197, 198, 199, 200, 201, 202, 203, 204, 205, 206, 208, 209, 210, 211
    variable 202, 203, 204, 209
out of balance moment 221, 222, 224, 227, 229, 238, 248, 249, 251, 254, 255, 258, 261, 263, 267, 300

P

PIMP 14, 17, 21, 22, 23, 24, 25, 26, 27, 34, 55
piston 117, 150, 151, 158, 163, 168, 176, 177, 178, 180, 184, 188, 191, 192, 194, 195, 196, 198, 199, 200, 204, 209, 213, 214, 215, 218, 219, 220, 222, 238, 241, 242, 248, 249, 250, 251, 254, 258, 261
piston
    rod 117, 184, 213, 219, 241, 254

pitch
circle 281
  cone 319
line velocity 285
radius 280, 292, 299, 309, 326
planetary carrier 315
plasticity 55
plate 121, 122, 124, 125, 127, 128
plating 2
Poissons Ratio 8, 129
polygroove plain section 79
power depression 248, 260
power elevation 223, 247, 248, 258, 276, 278, 280
pressure angle 281
pressure ratio 20
pre-stressing 32, 50
primer 109, 110, 118, 120
principal stresses 6, 7, 10, 11
propellant 2, 3, 4, 5, 17, 20, 34, 57, 68, 69, 81, 83, 108, 109, 110, 114, 120, 125, 126, 128, 129, 131, 132, 147, 148, 153, 155, 160, 166, 171, 279, 295, 298, 317
pure shear 45

R

radial stress 17, 18, 29, 31, 39, 40, 41, 55, 56, 123, 128, 130, 216
rate of fire 2, 4, 13, 91, 148, 152, 158, 182, 212, 333
recoil 3, 147, 148, 149, 150, 151, 152, 154, 155, 156, 161, 162, 164, 165, 168, 178, 181, 182, 184, 185, 186, 188, 189, 191, 208, 212, 213, 214, 215
  brake 149, 150, 151, 152, 153, 159, 161, 162, 163, 164, 165, 168, 176, 178, 181, 182, 183, 184, 186, 187, 188, 191, 192, 194, 197, 198, 200, 213, 214, 215, 217, 295, 298
    cycle 148,158,211,212
    force 3,17,162,163,179
    length 148, 154, 155, 157, 166, 177, 191, 218, 219
    system 16, 147, 148, 149, 150, 151, 152, 165, 166, 168, 190, 213, 241, 279, 312, 317
    systems 3, 149, 150, 152, 213

recoiling parts 132, 147, 148, 149, 151, 152, 153, 154, 155, 156, 157, 159, 160, 166, 176, 177, 178, 179, 180, 181, 182, 183, 187, 191, 197, 204, 218
recovery 82
recuperator 109, 148, 149, 150, 151, 152, 153, 157, 158, 159, 160, 161, 162, 165, 168, 176, 177, 178, 182, 183, 187, 188, 191, 192, 194, 195, 196, 213, 214, 218, 219, 220
rifling
  curve 65, 66, 67, 73
  profile 65, 73, 74, 79
rigidity 3, 92, 124, 128, 129, 288
RMP 13
roller 113, 116, 132

S

safety factor SF 2, 14, 16, 26, 27, 28, 38, 52, 216, 265, 270
seal friction force 152,153,159,161,162,176,177, 178,188,192,193,194,195,196,204,207,241,2 42,248,249,254,255,258,261
sealing 108, 109, 213
self-locking 277, 294, 295
sensitivity 275
spring
  alloys 271
  index 264, 269
  rate 224
  stiffness 226, 230, 233, 234, 235, 264, 268
stabilizers   278
strain energy   7, 8, 9, 11, 96
stress
  analysis 17, 19, 121, 301
  hoop 17, 18, 33, 34, 39, 41, 42, 56, 58, 84, 123, 128, 130, 216
  radial 17, 18, 31, 39, 40, 41, 123, 128, 130, 216
stripping 73
swaging 3

T

T slot 112, 113, 132
tangential strain 44, 46

tensile strength 63
theories of failure 6
thermal jacketing 5
thermal stresses 2
throttling 148, 150, 151, 162, 164, 183, 184, 185, 186, 191, 192, 193, 194, 195, 196, 197, 198
tolerances 16
torque 68, 144, 146, 243, 244, 245, 246, 254, 255, 266, 267, 272, 273, 279, 280, 295, 298, 300, 301, 308, 317, 318, 323, 331, 333, 334, 340
Trendline 234, 236, 269
Tresca 6, 7, 13
trigger 118
trunnion 222, 229, 243, 244, 245, 246, 275, 276, 279, 280, 295, 298, 300
twist
  combined 65
  increasing 65
  uniform 65, 67

U

uniform twist See twist
unsupported 35

V

variable orifice 200
vibration 91, 98, 100
volumetric strain 7, 8, 10, 11, 44

W

wall ratio 14, 18, 19, 26, 32, 33, 35, 36, 37, 38, 43, 52, 54, 63, 64, 83, 84, 86, 216
wear capacity 282, 287
Wear Factor 287
Welin breech screw 114
windage 65
working stresses 3, 58
worm 277, 287, 289, 290, 296, 297, 324, 325, 328
  gear see gear
  pair 277, 289, 294, 315, 326, 327

Y

yield point 3, 7, 14, 84, 216
yield strength 2, 4, 32, 33, 34, 38, 48, 51, 52, 61, 62, 63, 129, 213, 215
Youngs Modulus 8, 83